U0381654

Coal Preparation

选煤工程

(美) 颜芬华　著
Felicia F. Peng

·北京·
·Beijing·

Brief Introduction

This book is an edited version of Felicia F. Peng's lecture notes for the course MINE 427 Coal Preparation for undergraduate mining engineering students at West Virginia University from 1997 to 2016.

This book covers nearly all topics of operations after the coal has been brought up to the surface, from coal's environmental effects including global warming, coal's occurrence and composition and its effects on steel making and coal cleaning, various coal cleaning processes and their unit operations, to finally reclamation including black water treatment and equipment design as well as impoundment design. It comprises fourteen chapters and serves as a compilation of lecture materials spanning Professor Felicia F. Peng's teaching career. To enhance instructional effectiveness further, Chapter 14 lists seven sets of homework, five exams, and five laboratory experiments, it also offers very detailed instructions regarding objectives, safety equipments, experimental procedures, methods of data analysis, result presentations, and report formats.

The book is specifically designed for professionals in the coal preparation industry, including coal production and related technical personnel, research and development personnel, management personnel, as well as teachers and students specializing in coal preparation.

图书在版编目（CIP）数据

选煤工程＝Coal Preparation：英文/（美）颜芬华（Felicia F. Peng）著. —北京：化学工业出版社，2024.5

ISBN 978-7-122-45559-8

Ⅰ.①选…　Ⅱ.①颜…　Ⅲ.①选煤-英文　Ⅳ.①TD94

中国国家版本馆CIP数据核字（2024）第088969号

责任编辑：刘丽宏　　装帧设计：刘丽华

责任校对：宋　玮

出版发行：化学工业出版社（北京市东城区青年湖南街13号　邮政编码100011）

印　　装：北京建宏印刷有限公司

710mm×1000mm　1/16　印张17¾　字数349千字　2024年6月北京第1版第1次印刷

购书咨询：010-64518888　　售后服务：010-64518899

网　　址：http://www.cip.com.cn

凡购买本书，如有缺损质量问题，本社销售中心负责调换。

定　　价：298.00元

Felicia F. Peng

Formerly Department of Mining Engineering
West Virginia University
Morgantown, WV
USA

Formerly Department of Mineral Processing Engineering
Henan Polytechnic University
Jiaozuo, China

Editors

Chapter 1, 6, 10, 13	Dr. Bo Yu, PE, President of Richland Mining Consulting LLC
Chapter 2, 7	Dr. James Gu, Senior Metallurgist, Arkema S.A. Dr. Changliang Shi, Henan Polytechnic University, China.
Chapter 3, 8	Dr. Yu Xiong, Lead Metallurgist, Albemarle Corporation Dr. Ji Fang, Henan Polytechnic University, China.
Chapter 4, 9	Dr. Wan Wang, PE, PMP, Senior Researcher, Mining Applications, Nouryon Dr. Jiao Ma, Henan Polytechnic University, China.
Chapter 5, 11	Dr. Tony Chen, R&D Engineer, Cleancarbon AS
Chapter 12, 14	Dr. Syd Peng, Academician of US National Academy of Engineering

Felicia F. Peng, my wife for 53 years, was born on October 22, 1937 in Keelung, Taiwan, China and passed away on June 10, 2021 in Morgantown, WV, USA. This book is in memory of her life.

Felicia loved chemical engineering and studied diligently to receive her BS, MS, Engineer, and PhD in chemical engineering from Waseda University (Tokyo, Japan), Stanford University, and West Virginia University, respectively.

In July 1974, she joined me to come to the School of Mines, West Virginia University in Morgantown, WV, she as a researcher with Coal Research Bureau, and me as an assistant professor with the Mining Engineering Program. In July 1975, she joined the newly established Mineral Processing Engineering Program (MPE) as assistant professor and taught mineral processing related courses and performed coal preparation related research. In 1997, Mineral Processing Engineering Department was terminated due to consistent low enrollment. She then chose to join the Mining Engineering Department for which I was chairman 1979-2006. Thereafter, she taught coal preparation to undergraduate mining engineering students and advanced coal preparation and mineral processing related courses to graduate mining engineering students, in addition to actively performing coal preparation related research until her retirement in 2016.

Felicia loved teaching. She had a strong passion for being associated and working with students. She spent countless time and never tired of chatting and mentoring students. For that reason, I think, that's why her mind always stayed young and lively in her life. She did not retire until after her 80th birthday! She was an excellent mentor and trusted friend for her students in all her teaching career.

Felicia was a chemical engineer by academic training. Her educational background reflects very well in this book. In several chapters, rather than employing traditional narrative statements as mining engineering students would prefer, she started with fundamentals of and engaged in exhaustive mathematical derivation of theories of fluid mechanics as applied to various coal cleaning processes, followed by example problems and solutions such that students would know how to apply the engineering principles. Another example is that her lists of chemicals used in coal cleaning processes are supplemented by material ion bond structures and chemical reaction equations.

This book is an edited version of Felicia's lecture notes for the course MINE 427 Coal Preparation for undergraduate mining engineering students at West Virginia University from 1997 to 2016. It had many versions as it evolved with her yearly offer of the course. This book has fourteen (14) chapters that are a compilation of her lecture notes over her teaching career. It covers nearly all topics of operations after the coal has been brought up to the surface, from coal's environmental effects including global warming, coal's occurrence and composition and their effects on steel making and coal cleaning, various coal cleaning processes and their unit operations, to finally reclamation including black water treatment and equipment design as well as impoundment design. AutoCad plotting of flowsheets is extensively used. One unique feature is that Chapter 14 lists seven pieces of homework, five exams, and five laboratory experiments. The laboratory experiments give very detailed instructions regarding objectives, safety equipments, experimental procedures, methods of data analysis, result presentations, and report formats. However, it must also be noted that this is a lecture note, many topical areas are presented in bullet points format without details.

This book is the result of teamwork of Felicia's former students, as listed in the previous page, who graduated from the Mining Engineering Department, West Virginia University, with Dr. Bo Yu as the team leader.

Thanks are also due to Qingqing Huang, Department of Mining Engineering, West Virginia University for reviewing the draft manuscript.

Syd S. Peng
Morgantown, WV, USA
September, 2023

Contents

INTRODUCTION

1.1 Coal Preparation is a Subsystem of Mining Operations

Coal preparation plants and surface materials handling facilities are two of the subsystems of mining operations. One component of the coal preparation plant, commonly known as the coal cleaning facility, provides the flexibility of preparing certain coal qualities required by the clients. The typical operation components and the material flow in the subsystem are shown in Figure 1.1. The major components of this subsystem consist of a run-of-mine (ROM) haulage and transportation system, a crushing plant, raw coal material handling facility, a coal preparation plant, refuse disposal and tailings storage facilities, clean coal product stockpiles and storage units, clean coal product material handling facility, and product transportation modes.

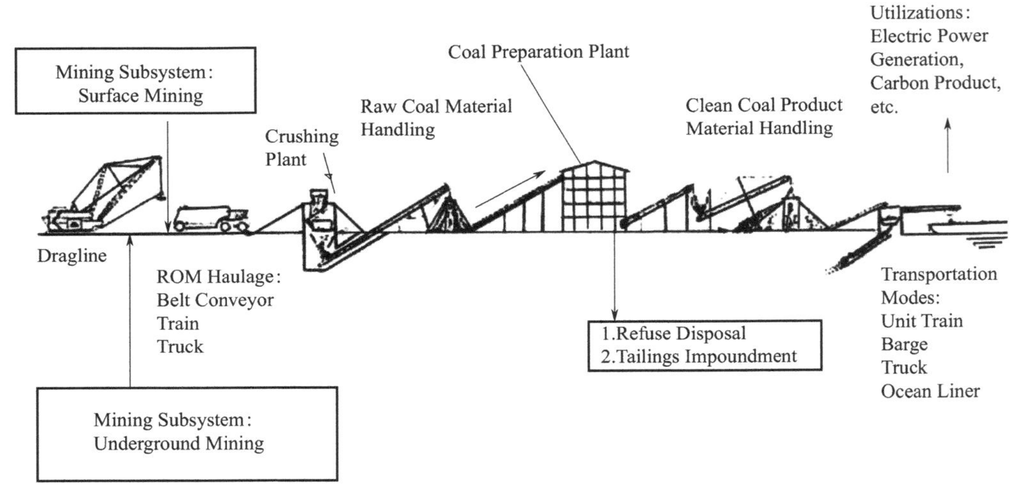

Figure 1.1 Subsystems of mining operations: coal preparation plant and materials handling facilities

1.2 Objectives of Coal Cleaning

Coal preparation is the removal of undesirable materials from the run-of-mine (ROM) coal by employing separation processes. The separation processes can differentiate between the physical and surface properties of coal and impurities. Through coal preparation, a pure uniform coal product is achieved. The selection of a proper separation process is based upon a laboratory study of the ROM coal, called the washability study. The goals of coal cleaning include:

① To produce uniformly sized and distributed coal, and to ensure high quality of clean coal products, to meet market specifications.

② To reduce incombustible mineral matters, minimize moisture content, and increase the heating value of clean coal products, for minimizing transportation cost.

③ To remove pyritic sulfur and other trace elements such as mercury (Hg) for air pollution control.

④ To produce quality metallurgical coal for improving coking properties.

⑤ Most importantly, to generate revenues for a coal mining operation.

1.3 Coal Cleaning Processes

Coal preparation includes wet or dry processes, using chemical or physical methods. The design of the plant flowsheet depends upon the characteristics of the coal being processed. Figure 1.2 shows a typical configuration of the coal preparation flowsheet. It consists of four subsystems: pre-treatment, cleaning, subsequent treatment, and products.

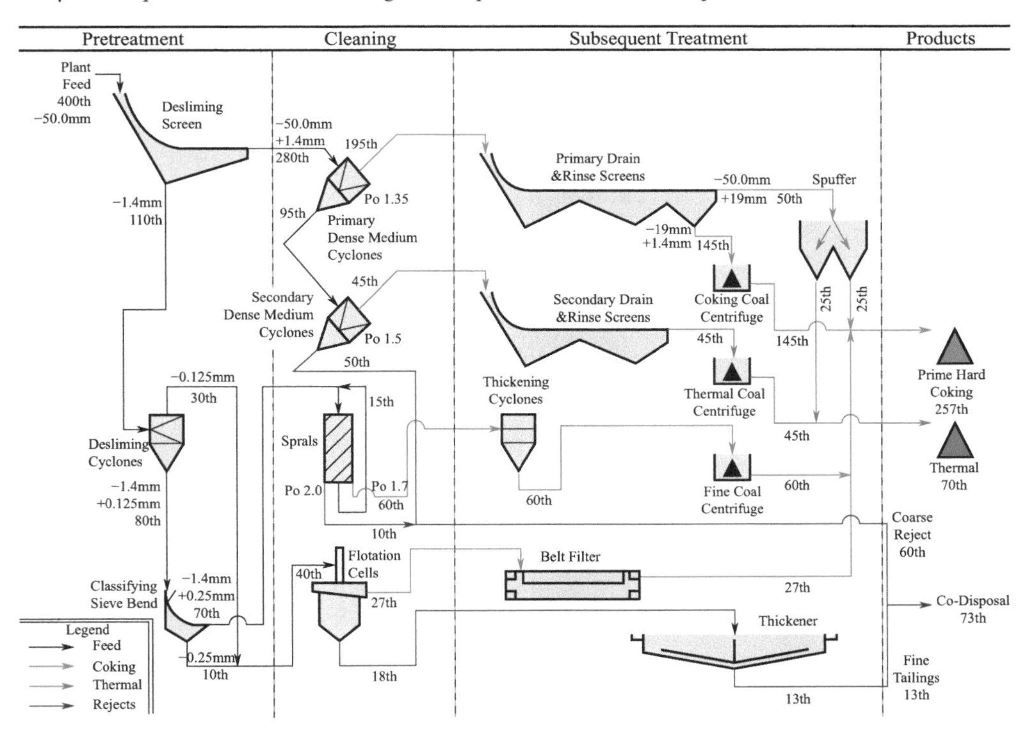

Figure 1.2 Typical components of coal preparation plant flowsheet

Most coal preparation plants use wet separation processes for higher separation efficiency, avoiding dust control problems when treating fine coal particles. The most commonly used separation processes are gravity-based separation and surface property-based flotation technologies. Another separation process is known as solid-solid particle separation, which is particle-size dependent and is also dependent on relative density or particle surface property. Thus, sizing coal feed prior to the separation processes becomes a necessary step.

Figure 1.3 shows an example of the mass distribution versus specific gravity and size in raw coal and clean coal. In this case, the components with specific gravity less than 1.40 remain. Conversely those with specific gravity larger than 1.40 are removed.

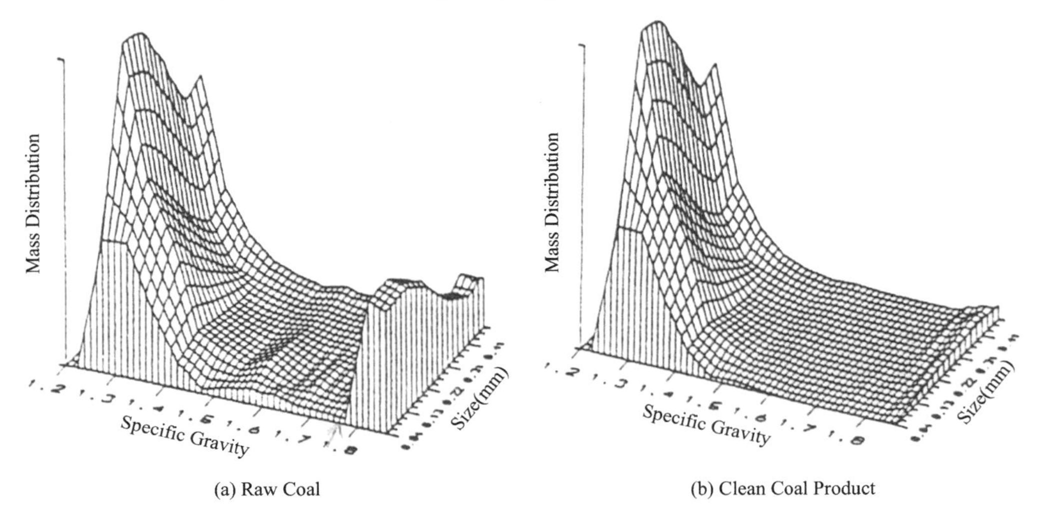

(a) Raw Coal (b) Clean Coal Product

Figure 1.3 Mass distribution versus specific gravity and size in raw coal and clean coal

1.4 Coal Utilization

Coal has various uses, for example, in year 2005 when U.S. coal consumption peaked, the dominant use of coal in the United States (92 percent of the total) was for electric power generation. Coke plants (2 percent) and other industrial uses (5 percent) accounted for most of the remainder (such as carbon and chemical sources), with a small amount of coal still used in residential and commercial buildings (such as non-smoke briquettes and logs)

① Electric power generation - Electricity generation and industry steam generation.

About 90 percent of the 313,000 megawatts (MW) of coal-fired generating capacity in the United States in 2005 was based on combustion of pulverized coal (National Academies Press, 2007). This process involves reducing the coal to a powder that is burned in a boiler to generate high-pressure, superheated steam that drives a turbine connected to an electric generator. The steam is then condensed back to a liquid and returned to the boiler to repeat the cycle.

② Steel making – Metallurgical coke for base metals and iron ore (pyro-metallurgical) processes.

The production of coke from coal is a centuries-old technology in which low-ash "metallurgical" coal is heated in an oven to drive off volatile matters, leaving a high-purity carbon product that is used in blast furnaces to produce iron for steel making. Modern coke plants consist of a battery of long, narrow, brick-lined rectangular ovens into which coal is fed. The volatile gases driven off by heating are collected, cleaned, and used as fuel. The hot coke product is pushed out of the oven into a rail car, quenched with water to cool it, and then shipped for use in steel making.

③ Gasification and Liquefaction.

Coal liquefaction technology has long been used to produce high-quality transportation fuels, most notably in South Africa, which boasts the largest commercial facility in the world (the SASOL Group). Substitute natural gas (SNG1) also can be produced from coal, and one commercial plant has been operating in the United States since the 1980s. In both cases, coal gasification is a key technology. By adjusting the ratio of carbon monoxide and hydrogen in the syngas product, either gaseous or liquid products can be manufactured with the proper choice of catalysts and operating conditions.

1.5 Environmental Considerations

1.5.1 Clean Air Act of 1990

Burning coal at power plants creates emissions of sulfur dioxide (SO_2), nitrogen oxides (NO_X), particulate matter (PM), carbon dioxide (CO_2), mercury (Hg), and other pollutants. NO_X and SO_2 emissions contribute to the formation of ground-level ozone and fine PM, which can lead to respiratory and cardiovascular problems, and exposure to mercury can increase the possibility of health issues ranging from cancer to immune system damage. The power sector has significantly reduced many of these pollutants over the past two decades, but important health and environmental concerns persist. The Clean Air Act of 1990 stipulates the maximum limits allowable and deadlines for compliance for those pollutants.

The emission rate of SO_2 and Hg from coal-fired power plants for burning certain quality of coal as fuels are determined as follows:

Example 1: A coal has 2% sulfur content and heating content of 13,000 Btu/lb. Calculate the sulfur dioxide per million Btu (MMBtu) emission rate.

Solution: Calculate the molecular weight of SO_2

since $S + O_2 = SO_2$,

molecular weight: $32 + 16 \times 2 = 64$.

S in coal will turn into SO_2 after burning; therefore, burning 1 lb of coal will generate SO_2 emission:

$1 \text{ lb} \times 2\% \times 64 / 32 = 0.04 \text{ lb}$

Therefore, the sulfur dioxide emission rate will be:

$0.04 \text{ lb} / 13000 \text{ (Btu/lb of coal)} \times 10^6 = 3.08 \text{ (lb } SO_2 \text{ /MMBtu)}$

Example 2: A coal with a heat content of 13000 Btu per lb has 0.23 ppm Hg per lb, calculate the mercury emission rate in lb Hg per trillion Btu.

Solution: For 1 lb coal, the mercury content is $1 \text{ lb} \times 0.23 \text{ ppm Hg/lb} = 0.23 \text{ lb Hg}/10^6$.

In order to generate 1 trillion Btu, it needs to burn $1 \times 10^{12}/13000 \text{ Btu/lb} = 76.9 \times 10^6 \text{ lb}$ coal.

Therefore, the mercury emission rate is $0.23 \text{ lb Hg}/10^6 \times 76.9 \times 10^6 = 17.69 \text{ lb Hg}$

1.5.2 Carbon Dioxide Emission

Carbon dioxide is known as a "greenhouse gas," which can trap heat in the earth's atmosphere, preventing it from escaping into space and keeping the earth warm (Figure 1.4). The principal anthropogenic sources of CO_2 are the combustion of fossil fuels, cement manufacturing, and deforestation. In 2005, fossil fuels supplied over 85% of the global energy demand. Fossil fuel usage is not likely to cease any time soon, either in industrialized or less-developed countries. In the 21st century, this situation will continue for decades.

The effect that anthropogenic emissions of CO_2 have on global climate has been of increasing public concern. In 1992, 143 nations signed the United Nations Framework Convention on Climate Change and in 1997 the Kyoto Protocol, which created legal commitments for reduction in emissions of greenhouse gases. Therefore, a variety of strategies are needed to reduce CO_2 emissions and remove carbon from the atmosphere in order to mitigate the potential effects of climate change. The disposal of CO_2 is becoming the focus of many research attempts.

The ocean covers more than 60% of the earth's surface and plays an important role in the global ecosystem. The global carbon cycle shows that the ocean takes up 2.2 Giga tons of Carbon (GtC) of average annual anthropogenic CO_2 emissions, which are approximately 7.4 GtC annually. Nevertheless, the carbon flux between the atmosphere and ocean is even over an order of magnitude larger than the net carbon exchange. This 2.2 GtC per year is accompanied by 92.2 GtC and 90 GtC fluxes into and out of the ocean. In actuality, the oceanic waters already contain an estimated 40,000 GtC compared with 750 GtC in the atmosphere and 2,200 GtC in the terrestrial biosphere. Therefore, if measures could be taken to alter the gross CO_2 exchange process, ocean and terrestrial systems would become the largest potential sink to sequester the anthropogenic CO_2 emissions. There are three approaches to carbon management.

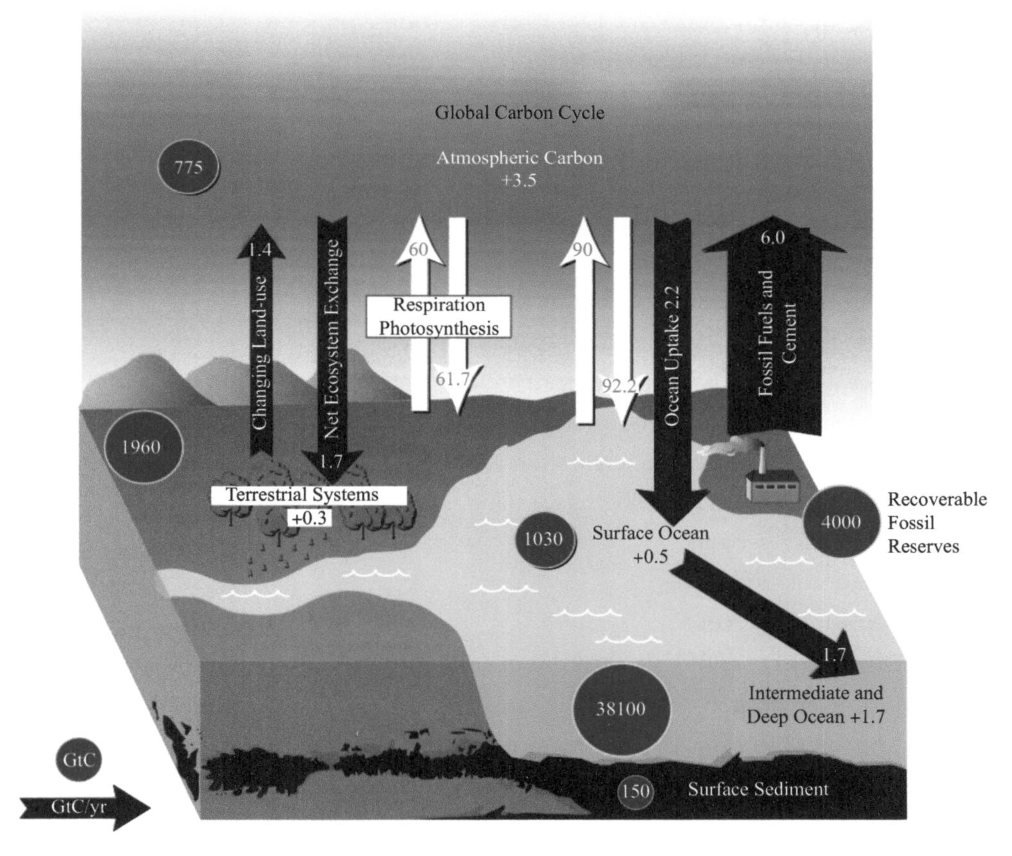

Figure 1.4 Human-induced changes in the global carbon cycle resulting from increases in the combustion of fossil fuels and changing land-use patterns

(Note: Solid arrows indicate the average magnitude of perturbation in carbon fluxes and the fate of carbon resulting from these activities averaged for the first half of the 1990s. Net fluxes (black arrows) and gross fluxes (gray arrows) are in billions of tonnes of carbon per year. Annual net additions of carbon (shown as + numbers) to the atmosphere, ocean subsystems, and terrestrial systems from anthropogenic sources are in billions of tonnes of carbon per year. Pool sizes (circles) are shown in billions of tonnes of carbon.

① The first approach is to increase the efficiency of primary energy conversion and end use so that fewer units of primary fossil energy are required to provide the same energy service.

② The second approach is to substitute lower-carbon or carbon-free energy sources for current sources. This strategy might involve, for example, substituting low-carbon fossil fuels such as gas for coal and oil, using renewable energy supplies such as solar, wind or biomass, or increasing the use of nuclear power.

③ Carbon sequestration, the third approach, refers to the removal of carbon dioxide (CO_2) from either manmade emissions or the atmosphere, and the safe, essentially permanent storage of CO_2 or other carbon compounds. The objective can be achieved by keeping the carbon emissions from reaching the atmosphere through capture and secure storage, or by removing carbon from the atmosphere through various means and fixing it.

In 2005, roughly one third of the United States' carbon emissions came from power plants and other large point sources such as cement manufacturing. To stabilize and ultimately reduce concentrations of this greenhouse gas, it will be necessary to employ carbon sequestration—carbon capture, separation and storage or reuse.

① Cost effective CO_2 capture and separation processes.

a. Absorption (chemical and physical) and adsorption (physical and chemical)

b. Low-temperature distillation and gas separation membranes

c. Mineralization and biomineralization

② CO_2 sequestration in geological formations including,

a. Oil and gas reservoirs

b. Unmineable coal seams

c. Deep saline reservoirs

③ Direct injection of CO_2 into the deep ocean and stimulation of phytoplankton growth via iron fertilization.

④ Improved full life-cycle carbon uptake of terrestrial ecosystem, forest lands, agricultural lands, biomass croplands, deserts, and degraded lands, boreal wetlands and peatlands.

⑤ Advanced chemical, biological, and decarbonization concepts, including biological systems, advanced catalysts for CO_2 or CO conversion, novel solvents, sorbents, membranes and thin films for gas separation, engineered photosynthesis systems, non-photosynthetic mechanisms for CO_2 fixation (methanogenesis and acetogenesis), genetic manipulation of agricultural and trees to enhance CO_2 sequestering potential, advanced decarbonization systems, and biomimetic systems.

1.6 Transportation Modes of Coal

Like all fuels, coal must be transported to an end user before it can be used. Specific transportation needs vary. The Gulf Coast lignite is generally transported over very short distances to mine mouth power plants. The Appalachian and Illinois Basin coals are typically transported over somewhat longer distances from mine to market, and coal mined in the Powder River Basin may travel distances ranging from less than 100 miles to more than 1,500 miles before it reaches the users.

1.6.1 Coal Distribution Methods

The five most common methods used to deliver coal to the customers are highway truck, river/tidewater barge, unit train rail system, the Great Lakes/ocean ships, and overland belt conveyor (Figure 1.5). The statistical data for those transportation modes are shown in Table1.1.

(a) Unit train rail system

(b) Highway coal truck

(c) River barge

(d) Great lakes and ocean cargo ships

(e) Overland belt conveyor

Figure 1.5 Five most common methods of coal transportation in USA

Table 1.1 Distribution of coal by transportation method—1990—2020 (thousand short tons)

Year	Railroad	Truck	River /Tidewater	G. Lakes	Tramway / Conveyor
1990	528,579	105,281	142,982	12,983	123,498
1991	512,443	98,183	138,444	11,494	119,508
1992	512,969	104,239	140,880	13,039	119,362
1993	528,736	115,555	128,885	11,250	95,188
1994	582,807	114,888	135,677	12,001	100,640
1995	589,353	104,076	132,786	12,096	100,002
1996	609,657	99,352	144,965	13,377	98,934
1997	614,892	122,188	160,794	13,919	96,128
1998	648,382	123,000	147,017	19,675	102,749
1999	676,691	112,083	137,191	13,745	100,332
2000	660,988	109,932	138,499	12,309	92,992
2001	707,074	120,295	129,430	10,558	97,343
2002	685,086	138,222	116,212	10,658	99,986
2003	680,523	128,480	102,993	11,091	115,262
2004	684,249	128,900	87,041	11,128	125,024
2005	679,818	107,989	88,468	10,404	84,977
2006	799,463	122,538	105,534	8,946	77,983
2007	790,413	122,229	111,589	7,883	73,735
2008	762,636	179,349	93,516	1,402	53,442
2009	689,161	150,775	99,241	1,568	48,626
2010	705,429	117,569	113,213	1,660	66,411
2011	705,843	102,063	104,350	1,745	59,487
2012	610,656	101,716	101,603	1,325	61,984
2013	582,641	98,323	108,200	337	68,824
2014	609,567	99,232	113,453	502	67,156
2015	560,039	85,246	96,944	394	72,861
2016	448,197	70,004	87,089	89	70,957
2017	459,892	62,484	81,244	83	67,088
2018	432,495	75,208	75,208	105	59,110
2019	408,474	55,754	72,930	90	53,818
2020	311,544	46,649	58,479	1	47,568

1.6.2 River Barge Transportation Systems

The major rivers in the United States used for distribution of coal are the three river systems and the Mississippi River system. There are three rivers, marine and rail systems and terminals. The three rivers consist of ① Monongahela River, ② Allegheny River, and ③ Ohio River.

COAL

2.1 Formation of Coal

Coal is derived from the massive accumulation of land-based plant life, mainly trees. This organic matter was deposited in sedimentary basins on land where the water was shallow. These basins were either close to the sea, often in the form of large lagoons, or inland, in the form of lakes or marshes. As a result of climatic variations, for example an increasingly heavy annual rainfall, it is thought that enormous forests sank below water, and their debris accumulated in sedimentary basins where it was rapidly covered by large quantities of mud and sand. Figure 2.1 illustrates the process of coal formation.

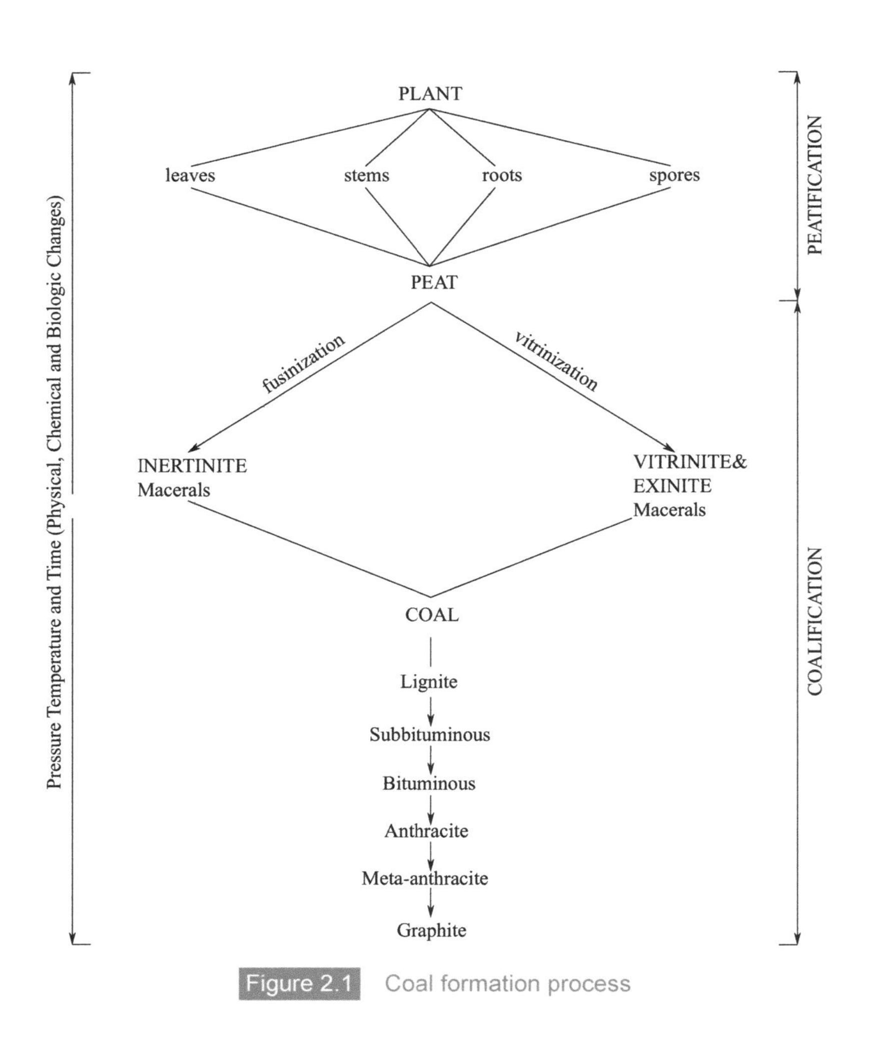

Figure 2.1 Coal formation process

In the maritime areas, a sudden subsiding of the basin resulted in an inflow of seawater, decimating the forest. After these catastrophic episodes, the forest grew again until a new disaster happened. This process has been repeated many times throughout geological time. This repetitive cycle of events explains why, in the substrata, layers of coal alternate with layers of clay or sandstone (compacted sand). Subsequently, the sedimentary basin sinks gradually under the weight of the sediments and the layers of dead plant matter are slowly transformed under conditions of increasing temperature and pressure.

The cellulose in the wood is first of all changed into humic acids, then into bitumens and finally into elementary carbon. The process is extremely long: the oldest coal has the highest carbon content (90 to 95%). The most favorable period for the creation of coal was the Carboniferous Period (carboniferous means "coal-bearing") between 360 and 290 million years ago.

However, smaller quantities of coal continued to be formed in certain regions during all the subsequent epochs: the Permian Period (290 to 250 million years ago), and the Secondary Period (Mesozoic Era, 250 to 65 million years ago). The vegetal masses buried in the Tertiary Period (Cenozoic Era), less than 65 million years ago, are generally less evolved: they are often lignite deposits which still contain a lot of volatile matter (bitumens and residual lignite), but whose carbon content is lower.

Coals of excellent quality can be found dating from the Tertiary Period, brought into early maturity by the heating effects of colliding tectonic plates: Paleocene coal (65 to 55 million years ago) from Colombia and Venezuela, or Miocene coal (20 million years ago) from Indonesia.

Finally, recent deposits (from 10,000 years ago until the present time) are very rich in fibrous debris. These consist of peat and they do not contain any elementary carbon - they are not sufficiently buried. Peat is found in peat marshes, boggy ground where it was plant matter such as mosses (sphagnum or peat moss) and grasses rather than the tree remains of forests that formed the deposits.

2.2 Representative Structure of Chemical Groups in Bituminous Coal

Coal is a complex hydrocarbon. It is about two-thirds inorganic carbon and one-third organic carbon compounds with traces inorganic minerals. The representative structure of the chemical group of bituminous coal is shown in Figure 2.2.

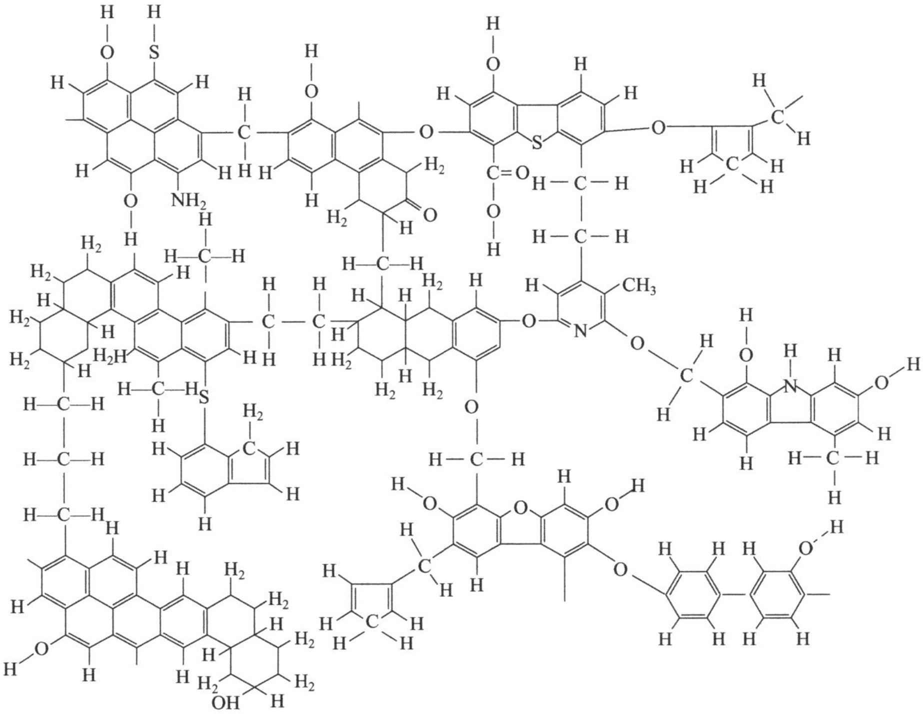

Figure 2.2 A Representative structure of the chemical groups in a bituminous coal

2.3 Parr Formula and Classification of Coal by Rank

Classification of coal by rank as shown in Table 2.1 (ASTM, D388 1998) is according to their degree of metamorphism, or progressive alternation in the natural series from lignite to anthracite. Classification is according to fixed carbon and gross calorific value calculated to the mineral-matter-free basis in accordance with the Parr formula. The simplest method for determining the amount of mineral matter present in a coal is to determine the ash and sulfur contents and to make corrections for the changes taking place in these components during combustion. The Parr formula are given below. Figures 2.3 and 2.4 show the inherent moisture, volatile matter and fixed carbon of a coal by ranks and calorific value of coal by rank, respectively. Figure 2.5 shows the basis of coal classification by rank in U.S.A.

The Parr formula:

$$\text{Total inorganic matter} = M + 1.08\ \omega_A + 0.55\ \omega_S \tag{2-1}$$

$$\text{Mineral matter (mm)} = 1.08\ \omega_A + 0.55\ \omega_S \tag{2-2}$$

$$\text{Dry, mm free FC} = (FC - 0.15\ \omega_S) \times 100\% / [100 - (M + 1.08\ \omega_A + 0.55\ \omega_S)] \tag{2-3a}$$

Table 2.1 Classification of coals by rank (ASTM D388 1998)

Class	Group	Fixed Carbon Limits,% (Dry, Mineral-Matter-Free Basis)		Volatile Matter Limits,% (Dry, Mineral-Matter-Free Basis)		Gross Calorific Value Limits, Btu/lb (Moist, Mineral-Matter-Free Basis)		Agglomerating Character
		Equal or Greater Than	Less Than	Greater Than	Equal or Less Than	Equal or Greater Than	Less Than	
I. Anthracitic	① Meta-anthracite	98	…	…	2	…	…	Non-agglomerating
	② Anthracite	92	98	2	8	…	…	
	③ Semianthracite	86	92	8	14	…	…	
II. Bituminous	① Low-volatile bituminous coal	78	86	14	22	…	…	Commonly agglomerating
	② Medium-volatile bituminous coal	69	78	22	31	…	…	
	③ High-volatile A bituminous coal	…	69	31	…	14000		
	④ High-volatile B bituminous coal	…	…	…	…	13000	14000	
	⑤ High-volatile C bituminous coal	…	…	…	…	11500	13000	
						10500	11500	agglomerating
III. Subbituminous	① Subbituminous A coal	…	…	…	…	10500	11500	non-agglomerating
	② Subbituminous B coal	…	…	…	…	9500	10500	
	③ Subbituminous C coal	…	…	…	…	8300	9500	
IV. Lignitic	① Lignite A	…	…	…	…	6300	8300	
	② Lignite B	…	…	…	…	…	6300	

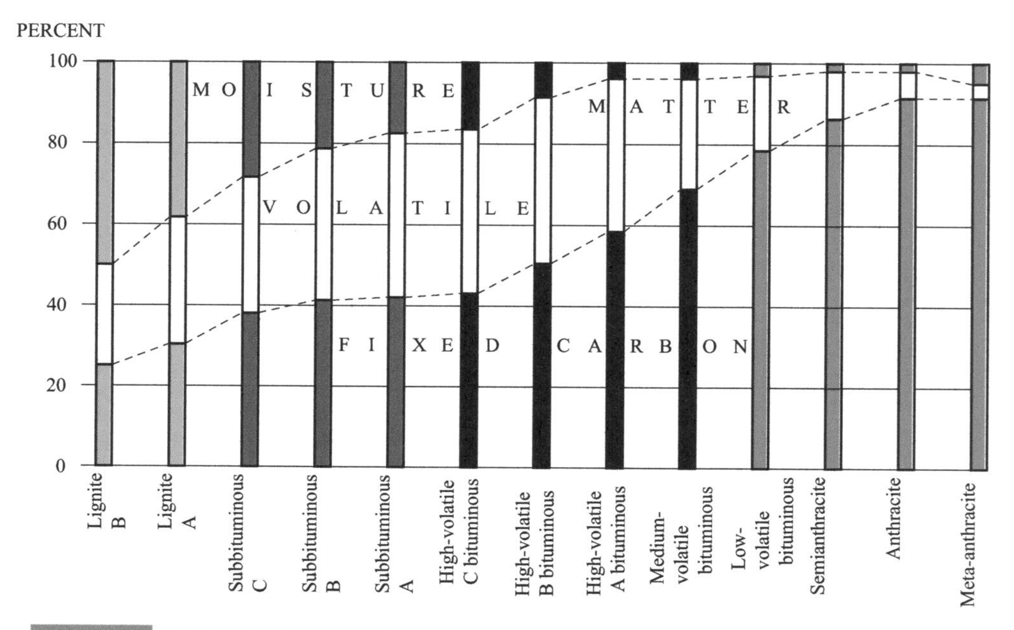

Figure 2.3 Natural inherent moisture, volatile matter and fixed carbon of coals by rank (Tully, 1996)

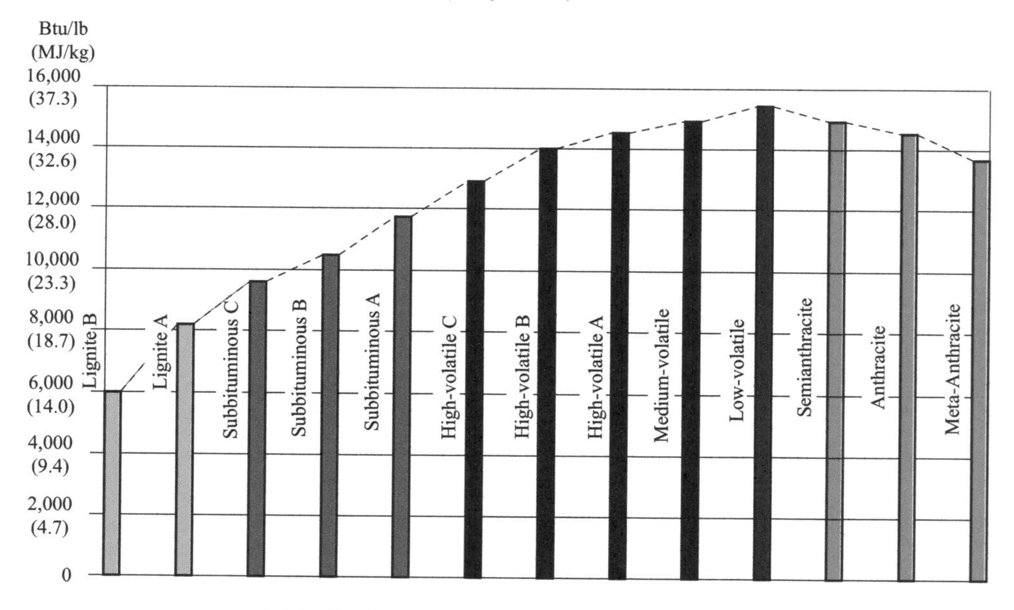

Figure 2.4 Calorific values of coals by rank

$$\text{Dry, mm free VM} = 100 - \text{Dry, mm free FC \%} \tag{2-4a}$$

$$\text{Moisture, mm free Btu} = (\text{Btu} - 50\ \omega_S) \times 100 \text{ per lb} / [100 - (1.08\ \omega_A + 0.55\ \omega_S)] \tag{2-5a}$$

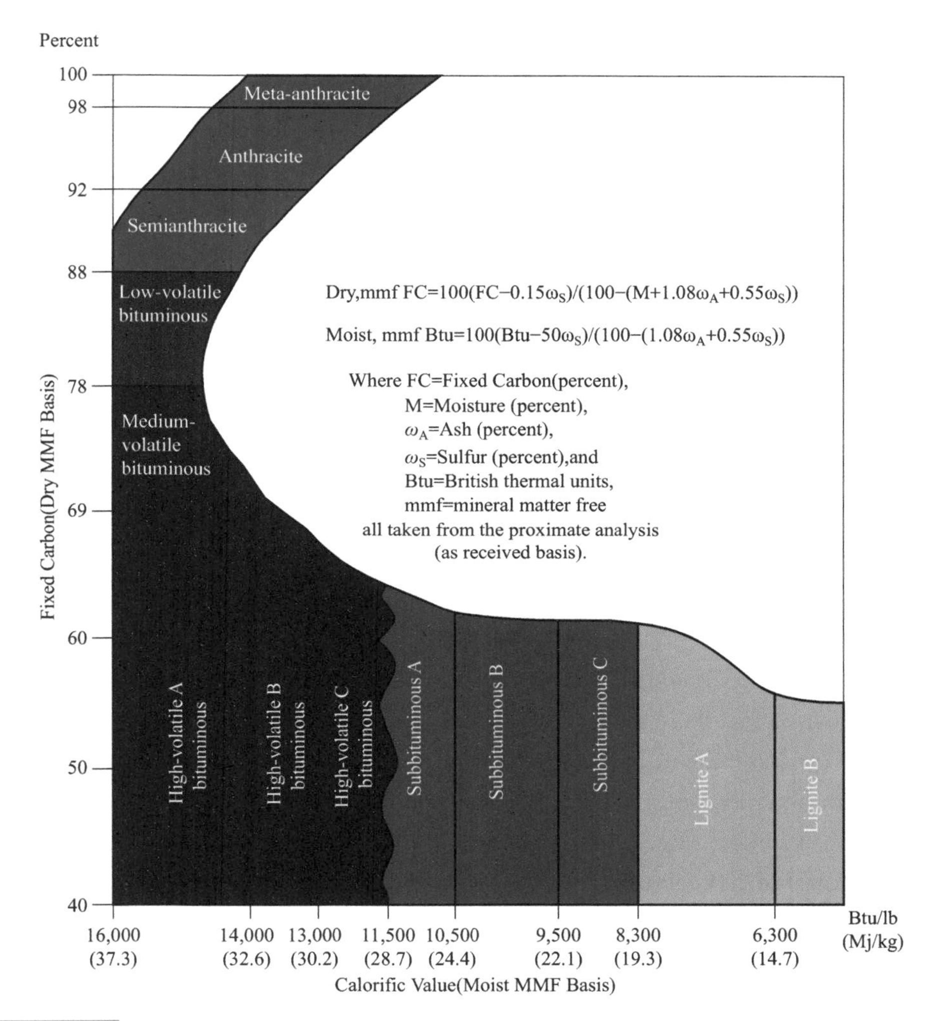

Figure 2.5 Basis for classification of rank of coals in the United States, and the formula used in making approximate rank determinations

Then, the following approximate formula can be derived:

$$\text{Dry, mm free FC} = (\text{FC}) \times 100\% / [100 - (\text{M} + 1.1\,\omega_A + 0.1\,\omega_S)] \tag{2-3b}$$

$$\text{Dry, mm free VM} = 100 - (\text{Dry, mm free FC}) \tag{2-4b}$$

$$\text{Moisture, mm free Btu} = (\text{Btu}) \times 100 \text{ per lb} / [100 - (1.1\,\omega_A + 0.1\,\omega_S)] \tag{2-5b}$$

where ω_A = ash, %;

Btu = British thermal units per pound, gross calorific value, Btu/lb;

FC = Fixed Carbon % (the fixed carbon content of coal is the carbon found in the residue remaining after the volatile matter has been expelled. Fixed carbon is not determined directly but is the amount that remains after subtracting the moisture, ash, and volatile matter from an air-dried coal.);

mm = mineral matter, %

VM = Volatile Matter (volatile matter represents that component of coal, except for moisture, that is driven off at high temperature in the absence of air. This material is derived mainly from the organic fraction of the coal, but also minor amounts of elements from the mineral matter present.);

M = moisture, %;

ω_S = sulfur, %.

The above quantities are all in coal on its natural inherent moisture basis as described in ASTM D388 (1998).

Note: ω_A and ω_S represent the percentages of these substances found by analysis of coal.

2.4 Coal Resources in the United States and West Virginia

The United States is endowed with a vast amount of coal. The U.S. Geological Survey (USGS) estimated that there are nearly 4 trillion tons of total coal resources in the United States. However, this estimate has little practical significance because most of this coal cannot be mined economically using current mining practices. A more meaningful figure is the ~267 billion tons of Estimated Recoverable Reserves (ERR) that is the basis for the commonly reported estimate of the coal reserves in the United States.

2.4.1 Coal Resources Distribution in the United States

Coal fields in the United States consists of the following three regions.

① Eastern Region- North, Central and South Appalachian regions and PA anthracite region.

② Central Region- Eastern interior and Western interior regions, Mississippi region, Texas region

③ Western Region- Fort Union region, Powder River region, North Central region, Big Horn Basin region, Wind River Basin region, Green River Hamas Fork region, Uintah region, Denver Raton Mesa, Southwest Utah, San Juan River basin, and Centralis Chehalis.

The coal fields, and their sulfur %, and mercury lb/trillion, and Btu distribution are shown in Figure 2.6, respectively.

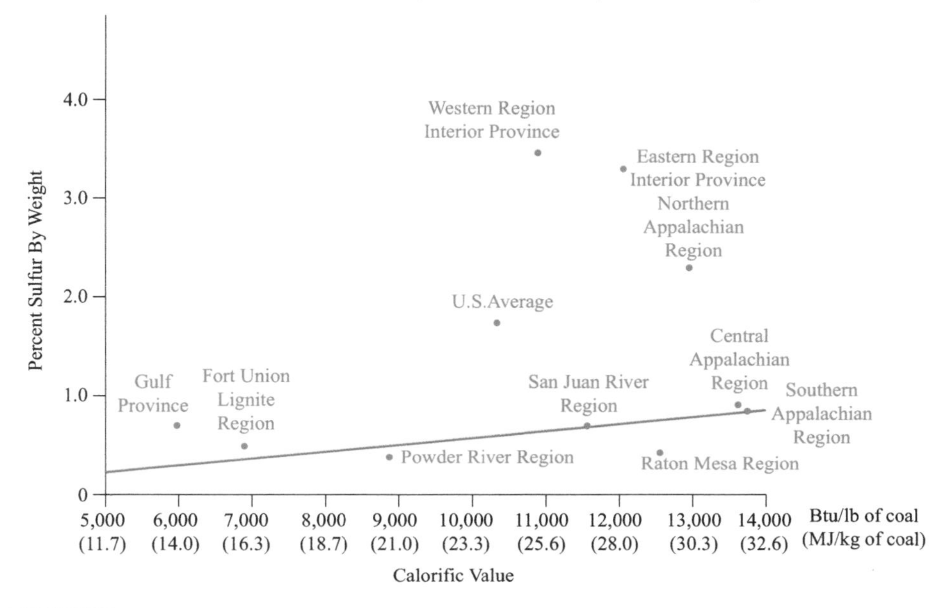

Figure 2.6 Mean sulfur and caloric values of raw coal samples by major coal fields in the United States

2.4.2 Coal Resources and Coal Seams in West Virginia

Coal has been an important part of West Virginia's economy for more than a century. The existence of coal in West Virginia was first reported in the early 1700s, but large-scale mining did not begin until the mid-1800s. In the 21st century, West Virginia is the second-largest coal producer in the U.S. after Wyoming. The state is also the largest producer of bituminous coal. Coal deposits underlie all but two counties, both of which are located in the state's eastern panhandle. Although coal occurs in 53 of the state's 55 counties, only 43 counties have economically recoverable reserves. All West Virginia coal is bituminous, but sulfur content varies across the state. The Central Appalachian region, which includes the southern part of West Virginia, is the nation's primary source of bituminous coal that is relatively low in sulfur. Coal from the Northern Appalachian region, which includes northern West Virginia, has relatively high sulfur contents. West Virginia is among the top three states in the nation in recoverable coal reserves at producing mines. Most of West Virginia's coal production and reserves are found in underground mines.

Table 2.2 is the coal seam sequence showing the 62 minable coal seams in West Virginia.

Table 2.2 Vertical distributions of 62 minable coal seams in West Virginia

Geological Unit	Minable Bituminous Coal Seams
Dunkard Groups	Washington
Monongahela Groups	Washington
	Waynesburg A
	Waynesburg
	Sewickley
	Redstone
	Pittsburgh
Conemaugh Groups	Little Pittsburgh
	Normantown
	Lower Hoffman
	Upper Clarysville
	Elk Lick
	Harlem
	Upper Bakerstown
	Bakerstown
	Brush Creek
	Mahoning
Allegheny Formation	Upper Freeport
	Lower Freeport
	Upper Kittanning
	Middle Kittanning
	Lower Kittanning
	Clarion
	No. 6 Block
	Upper No. 5 Block
	No. 5 Block
	Little No. 5 Block
	Stockton A
Kanawha Formation	Upper Mercer
	Stockton
	Coalburg
	Winifrede
	Chiton A
	Chilton
	Little Chilton
	Fire Clay

Geological Unit	Minable Bituminous Coal Seams
Kanawha Formation	Cedar Grove
	Williamson
	Peerless
	No. 2 Gas
	Powellton A
	Eagle
	Little Eagle
	Matewan
	Middle War Eagle
	Bens Creek
	Lower War Eagle
	Glenalum Tunnel
	Gilbert
	Douglas
	Lower Douglas
New River Formation	Laeger
	Sewell B
	Sewell A
	Sewell
	Welch
	Beckley
	Fire Creek
	Pocahontas No. 9
	Pocahontas No. 8
Pocahontas Formation	Pocahontas No. 7
	Pocahontas No. 6
	Pocahontas No. 5
	Pocahontas No. 4
	Pocahontas No. 3
	Pocahontas No. 2
	Pocahontas No. 1

COAL CHARACTERIZATION

3.1 Coal Sample Preparation, Analyses, and Specifications

Coal is a heterogeneous material consisting of various carbon-rich organic minerals (i.e., macerals), which contain impurities of a variety of inorganic minerals (e.g., Silica, Pyrite, Illite, etc.). Not only are there large differences in the properties and composition of coals in different seams but also in coal removed from different locations and elevations within a single seam. Since an effective and economical strategy to mine and process coal is based upon a coal's composition and properties, the development of reliable techniques to accurately characterize coal is critical. The sampling of coal includes two general steps: 1) sample acquisition and 2) sample preparation.

Sample acquisition consists of obtaining a representative sample. Sampling techniques vary according to the source of the sample. To suit all the needs of both exploration and production, the American Society of Testing Materials (ASTM) has consulted the expertise of engineers, chemists, and statisticians in developing standard methods for sampling coal from such diverse sources as deep mines, surface mines, stockpiles, conveyor belts, and railroad cars. In this section, it is assumed that this step, the acquisition of a gross sample, has been properly completed.

Sample preparation consists of systematically reducing the coal sample for size for chemical analysis without altering its properties. Since the ASTM standards for a moisture and an ash analysis require the use of one gram of minus 250 mm (-U.S. series No.60, −60 mesh Tyler) coal, and since a gross sample can consist of several tons of minus 6-in. material, the process of reducing a sample is necessarily tedious.

For situations where only a chemical analysis needs to be performed, the ASTM standard dictates that the entire gross sample be air-dried and reduced in size to minus No.4 U.S. series (−4 mesh Tyler) before successively reducing the size of the sample. The minus 4-mesh coal is then riffled until 1000 to 4000 grams remain. This coal is then pulverized to minus No.60 U.S. sieve (−250 μm, or −60 mesh Tyler) and subsequently riffled to obtain a 50-gram sample for chemical analysis. For obvious reasons, this method cannot be used when the size of coal is being studied. In that case, the size of the sample is reduced through the alternate shovel technique and cone and quartering as shown in Figure 3.1. The final four quarters might then be divided up for various size and other analyses. For cone and quarter techniques and equipment used for riffling see Figure 3.2.

- Using Long-pile Alternative Shovel Method and Cone and Quarter Method (ASTM D 346).

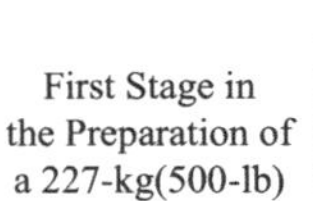

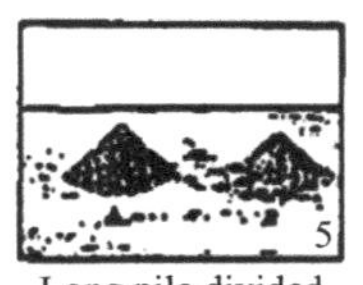

Figure 3.1 Standard hand method of crushing and reducing the gross sample quantity of coal

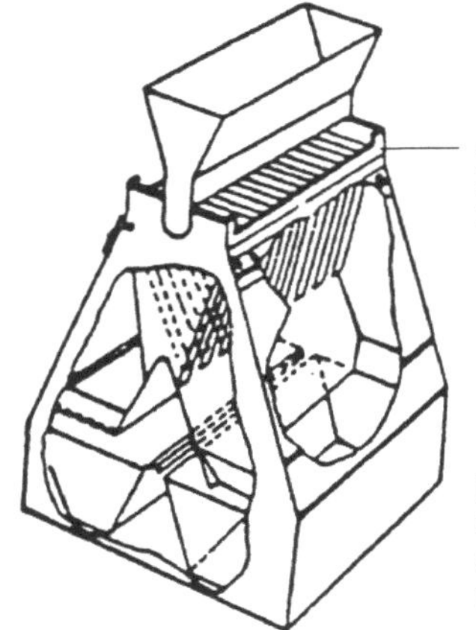

Figure 3.2 Riffle (left) and sample divider (right) for coal sampling

Many of the properties of coal vary with factors such as its composition and the presence of mineral matter. Different techniques have been developed to examine the properties of coal. These are X-ray diffraction, scanning and transmission electron microscopy, infrared spectrophotometry, mass spectroscopy, gas chromatography, thermal analysis, and electrical, thermal analysis, and electrical, optical, and magnetic measurements.

Knowing the physical properties of coal are important in the preparation and use of coal. For example, coal density ranges from about 1.1 to about 1.5 megagrams per cubic meter, or grams per cubic centimeter. Coal is slightly denser than water and significantly less dense than most rocks and mineral matter. Density differences make it possible to improve the quality of a coal by removing most of the rock matter and sulfide-rich particles through heavy liquid separation.

Other properties such as hardness, grindability, ash fusion temperature, and free swelling index (a visual measurement of the amount of swelling that occurs when a coal sample is heated in a closed crucible) can affect coal mining and preparation, as well as the way a coal is used. Hardness and grindability determine the types of equipment used for mining, crushing, and grinding, in addition to the amount of power consumed in their operations. Ash fusion temperature affects furnace design and operating conditions. The free swelling index provides preliminary information on the suitability of a coal for coke production.

The physical and chemical properties of coal are listed in Table 3.1.

Table 3.1 Chemical and physical properties of coal

Physical Properties	Organic Phases (Non-crystalline)	Mineral Phases (Crystalline)	Subjective Properties	Form of Sulfur
Size	Elemental analysis:	Silicates-clays and shales:	Proximate analysis:	Total sulfur
Shape	C,H,O,N, and	Oxides (SiO_2, Fe_2O_3)	Moisture, Ash, Volatile matter	Pyritic sulfur
Density	S-Inorganic		Fixed Carbon	Organic sulfur
Hardness				Sulfate sulfur
Particulate component	Unit Structure	Sulfide (FeS_2)	Rank	
	Lithobody	Sulfate ($CaSO_4$, Jarosite)	Calorific content	
Crystal	Lithotype: Marceral	Carbonates:	Free swelling index	
Pysto residual		($CaCO_3$, $FeCO_3$, Ankerite)		
	Vitrinite	Trace minerals:	Ash fusion temperatures	
	Pseudovitrinite	(Rutile, etc)		
	Fusinite		Hardgrove grindability index	
	Semifusinite	Trace impurities:		
	Micrinite:	(Mn, Zn, Cd, etc.)	Boney-slate	
	Granular Massive			
	Exinite:			
	Resinite, Alginite			
	Sporinite, Cutinite			

3.2 Qualities of Steam and Metallurgical Coals

Metallurgical coal has a low ash content, low moisture content and low sulfur and phosphorous contents. It is suitable to be converted into coke for steel making, whereas steam coal is mainly for generation of electric power. Tables 3.2 and 3.3 show examples of market specifications for steam and metallurgical clean coals for Pittsburgh seam coal.

Table 3.2 Example of steam coal quality specification

CHEMICAL PROPERTIES		Alkalies in Coal,%(D/B)		
		Na_2O	0.02	
Proximate Analysis		K_2O	0.09	
Moisture %	5.57			
Volatile Matter, % (A/R)	35.45			
Volatile Matter, % (D/B)	37.54	Water Soluble Alkalies		
Ash, % (A/R)	5.83	Na_2O	0.004	
Ash, % (D/B)	6.17	K_2O	0.001	
Fixed Carbon, % (A/R)	53.15			
Fixed Carbon, % (D/B)	56.29			
		PHYSICAL / Misc PROPERTIES		
Sulfur % (A/R)	1.40			
Sulfur % (D/B)	1.49	Hardgrove Grindability Index	53	
		Equilibrium Moisture,%	2.1	
Forms of Sulfur,%(D/B)		Free Swelling Index	8%	
Organic	0.98			
Pyritic	0.51			
Sulfate	< 0.01			
		Ash Fusion Temperatures	Reducing	Oxidizing
Calorific Value, BTU/lb.,(A/R)	13,405	Initial Deformation,*F	2,221	2,388
Calorific Value, BTU/lb.,(D/B)	14,196	Softening Temp.,*F	2,487	2,626
		Hemispherical Temp.,*F	2,522	2,675
Ultimate Analysis, %(D/B)		Fluid Temp.,*F	2,619	2,742
Carbon	79.82			
Hydrogen	5.23	Trace Elements in Coal, μg/gm (D/B)		
Nitrogen	1.52	Arsenic	3.12	
Chlorine	0.07	Beryllium	1.42	
Sulfur	1.49	Lead	3.45	
Ash	6.17	Fluorine	35.22	

Oxygen	5.70	Mercury	0.16
Ash Composition, % (In ash as ignited)		CALCULATED PARAMETERS	
SiO_2	50.33		
Al_2O_3	25.32	Sulfur Emission Ratios	
Fe_2O_3	14.59	Sulfur, lbs./mm BTU	1.05
TiO_2	1.03	SO_2, lbs./mm BTU	2.09
CaO	3.12		
MgO	0.60	Slag Characteristics	
Na_2O	0.38	T-250 Poise, *F	2,610
K_2O	1.42	Fouling Index	0.10
P_2O_5	0.43	Level	Low
SO_3	2.10	Slagging Index	0.39
Undetermined	0.67	Level	Low
		Base / Acid ratio	0.25
Phosphorus in Coal,%(D/B)	0.03	Silica Value	73.33

Table 3.3 Example of Metallurgical Coal Quality Specification

CHEMICAL PROPERTIES		RHEOLOGICAL PROPERTIES	
		Gieseler Plasticity	
Moisture, %	5.57	Max Fluidity, ddpm	29,988
		Max, Fluidity Temp,℃	431
Proximate Analysis, (% dry basis)		Initial Softening Temp, ℃	377
Volatile Matter	37.54	Solidification Temp, ℃	474
Ash	8.17	Plastic Range	97
Fixed Carbon	56.29		
		Amy Dilatation	
Sulfur (% dry basis)	1.49	Max. Contraction, %	−27
		Max. Dilatation, %	219
Oxidation Test (% Trans, at 17mm)	96	Initial Softening, ℃	362
		Initial Dilatation, ℃	412
Ash Composition, (% in ash as ignited)		Final Dilatation, ℃	458
SiO_2	50.33		
Al_2O_3	25.32	Sole Heated Oven	

Fe_2O_3		14.59
TiO_3		1.03
CaO		3.12
MgO		0.80
Na_2O		0.38
K_2O		1.42
P_2O_5		0.43
SO_3		2.10
Undetermined		0.68
Phosphorus in Coal,%(D/B)		0.03
Alkalies In Coal,%(D/B)	Na_2O	0.02
	K_2O	0.09
PHYSICAL / Misc PROPERTIES		
Hardgrove Grindability index		53
Coke Reactivity Test (calculated)		
Coke Reactivity Index (CRI)		36.9
Coke Strength After Reaction (CSR)		50.2
Ash Fusion Temperatures		Reducing
Initial Deformation, °F		2,221
Softening Temp., °F		2,457
Hemispherical Temp., °F		2,522
Fluid Temp., °F		2,619

Volume Change, % (52 BD, 2% H_2O)		−33.4
PETROGRAPHIC PROPERTIES		
Petrographic Analyses		
Maceral Composition, (vol.%)		
Reactives		
V-Types	7	2.2
	8	47.8
	9	22.5
Vitrinite		72.5
Exinite		5.6
Resinite		0.9
Semifusinite		1.7
Total Reactives		80.7
Inerts		
Semifusinite(2/3)		3.8
Micrinite		9.3
Fusinite		2.5
Mineral Matter		3.8
Total Inerts		19.3
Petrographic Indices		
Vit. Reflectance (R max), %		0.55
Comp. Balance Index		0.64
Rank/Strength Index		3.16
Calc. Stability Factor		36

3.3 Characterization of Coal for Combustion

Classification of coal by rank is according to its degree of metamorphism or progressive alteration, in the natural series from lignite to anthracite. Classification is according to fixed carbon and the gross calorific value calculated on the mineral-matter-free basis in accordance with the Parr formula.

3.3.1 Proximate and Ultimate Analyses

The proximate analysis test is widely used because it supplies a single integrated characterization of coal. This test includes the analysis of total moisture, volatile matter, and ash contents in coal sample and calculation of fixed carbon (ASTM D5142-90).

The Ultimate Analysis test is an elemental analysis, which covers the determination of moisture, carbon, hydrogen, nitrogen, sulfur, ash and oxygen (by difference) contents of a coal sample (ASTM D3176-89). The elemental analysis is conducted rather infrequently, except for sulfur analysis. Although technically a part of an ultimate analysis, sulfur is determined as frequently as the proximate analysis. Table 3.4 shows the typical proximate and ultimate analysis of Pittsburgh seam coal in West Virginia.

Table 3.4 Proximate and ultimate analysis of pittsburgh seam coal

Proximate Analysis		Ultimate Analysis	
Component	Weight, %	Component	weight, %
Moisture	2.5	Moisture	2.5
Volatile Matter	37.6	Carbon	75.0
Fixed Carbon	52.9	Hydrogen	5.0
Ash	7.0	Sulfur	2.3
Total	100	Nitrogen	1.5
		Oxygen	6.7
Heating Value (Btu/lb)	13,000	Ash	7.0
		Total	100

<u>Moisture of Coal</u>

The percentage of moisture present, commonly called "bed moisture", is more or less constant throughout a given mine and is a characteristic of coal rank. Such moisture ranges from 1, 2 or 3 percent in bituminous coal to 45 percent in lignite. Four types of moisture in coal need to be considered. They are:

① Inherent moisture in coal - moisture that exists as an integral part of the coal seam in its natural state, including water in pores but not that present in macroscopically visible features (as determined by Test Method ASTM D3173—87 or Classification ASTM D388).

Inherent moisture is considered equivalent to bed moisture but is not equated to the moisture remaining in a coal sample after air drying.

② Free moisture in coal - a portion of total moisture in coal (as determined by Test Method ASTM D3302) that is in excess of inherent moisture in coal and is referred to as surface moisture in coal.

③ Total moisture - that moisture determined as the loss in weight to the atmosphere under rigidly controlled conditions of temperature, time, and air-flow.

④ Residual moisture - that moisture remaining in the sample after having determined the air-dry loss.

For total moisture determination, prepare a coal sample to minus 8 mesh (−2.36 mm). Preheat the oven to 104~110℃. Weigh 1 gram of the coal sample and place the sample in the oven for 1 hour. Remove the sample from the oven and weigh immediately. Return it to the oven for an additional ½ hour, remove and weigh again. Repeat the drying at ½ hour intervals until the weight loss is no more than 0.05% for the ½ hour period (see ASTM D2961—87). An Ohaus moisture balance may be used as an alternative method.

Volatile Matter (VM)

Weigh 1 gram coal sample (−250 μm, -No.60) in a weighed platinum crucible, closed with a cover that fits closely enough so that the carbon deposit from bituminous coal does not burn away from the underside. Place on Nichrome-wire supports and insert directly into a preheated furnace chamber. The furnace temperature shall be maintained at (950 ± 20)℃. The coal sample shall be heated for exactly 6 minutes at 950℃ in the absence of air. The amount of weight lost equals to the volatile matter (ASTM D3175—89A).

Ash Content

Transfer 1 gram of coal sample (−250 μm) to a weighed capsule and cover it quickly. Place the capsule in a cold furnace and heat it gradually at a rate that the temperature reaches 450~500℃ in one hour. Continue heating to 700~750℃ by the end of the second hour. Continue to heat it for additional two hours. Remove the capsule from the muffle and place the capsule aside to cool in a desiccator. The amount left in the capsule is the ash content (ASTM D3174—00).

Fixed Carbon

The fixed carbon of the coal sample is calculated by using the following equation:

$$\text{Fixed Carbon \%} = 100\% - \text{Moisture\%} - \text{VM\%} - \text{Ash\%} \tag{3-1}$$

3.3.2 Sulfur Forms and Analyses (ASTM D3177—89, ASTM D2492—90)

Sulfur occurs in coal in three forms: organic sulfur, pyrite sulfur, and sulfate sulfur. They are described as follows:

Organic Sulfur

Organic Sulfur is chemically linked to the coal structure and therefore is very difficult to remove from the coal. The organic sulfur is present in one or more of the following forms:

① Mercaptan R-SH;

② Sulfide or Thio-ether R-S-R′;

③ Disulfide R-S-S-R;

④ Aromatic system containing the thiophene ring.

The organic sulfur in coal ranges from 20.8 to 97.1% of total sulfur and has a mean value of 51.2% of the total sulfur. The total sulfur in coal ranges from 0.46% to 4.97%, while pyrite and organic sulfurs range from 0.18% to 2.82%, and 0.45% to 2.40%, respectively. Organic sulfur is determined by the following equation:

$$\text{Organic Sulfur} = \text{Total Sulfur} - \text{Sulfate Sulfur} - \text{Pyritic Sulfur} \tag{3-2}$$

Pyrite Sulfur

Pyritic and marcasite sulfur have the same chemical composition as FeS, but they have different crystalline forms. Pyrite has an isometric orthorhombic form, while marcasite has a marcastic orthorhombic form. Both forms of sulfur are designated as pyrite. Pyritic sulfur is calculated as stoichiometric combination with iron.

Sulfate Sulfur

Sulfate sulfur is present in coal primarily as either gypsum or iron sulfate. The former is present in fresh coal and the latter results from the weathering of coal, although iron sulfates may be present in fresh coal in small amounts.

To determine the sulfate sulfur content in a coal sample, sulfate sulfur is extracted from the sample with diluted hydrochloric acid. The extracted sulfate sulfur is then precipitated as $BaSO_4$ and determined gravimetrically. Sulfates are soluble in hydrochloric acid, but pyritic sulfur and organic sulfurs are not. (ASTM D2492-90).

Total Sulfur

There are two alternative procedures for the determination of total sulfur in coal samples. They are:

① Eschka Method - A weighed sample is mixed with 1 part sodium carbonate and 2 parts light calcined magnesium carbonate (Eschka mixture) and ignited together. The sulfur compounds evolved during combustion react with the sodium carbonate and, under oxidizing conditions, are retained as sodium sulfate. The light magnesium carbonate reduces the bulk density and allows access of air to the mass. The sulfur is dissolved in hot water and then precipitated from the resulting solution as barium sulfate ($BaSO_4$). The precipitate is filtered, washed, and weighed.

② Bomb Washing Method (Figure 3.3) - Sulfur is precipitated as $BaSO_4$ from oxygen-bomb calorimeter washings, and the precipitate is filtered, washed, and weighed.

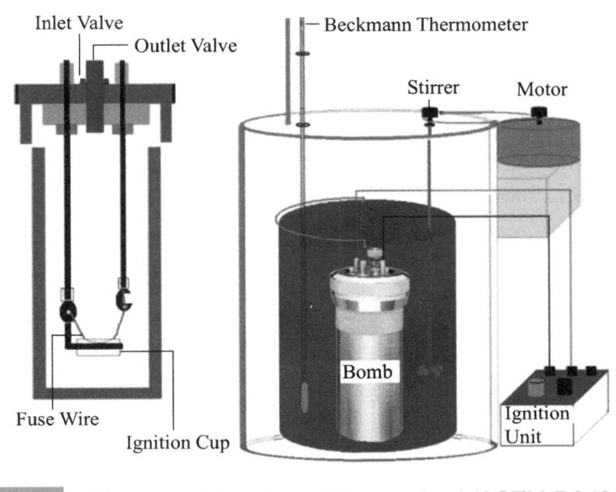

Figure 3.3 Diagram of bomb washing method (ASTM D2492—90)

3.3.3 Calorific Value of Coal - Adiabatic Bomb Calorimeter (ASTM D5865—99a)

The gross (or higher) calorific value of coal is determined by using a Parr calorific bomb. The products of combustion at the end of the determination are in the form of ash, liquid water, gaseous carbon dioxide, sulfur dioxide, and nitrogen. The net (or lower) heating value is calculated from the gross heating value and is defined as the heat produced at constant atmospheric pressure conditions such that the water in the combustion products remains in the form of vapor. The net heating value represents heat released under conditions of burning coal as fuel.

3.4 Hardgrove Grindability Index (HGI)

The Hardgrove Grindability Index is used to measure the hardness, strength, tenacity, fracture, and other properties of a given coal using a miniature grindability mill of the ring-and-ball type. The smaller the HGI, the harder the coal texture is and the less grindable the coal is.

Grindability is an important factor in the design of a coal mill. As grindability depends on many unknown factors, HGI is determined empirically using a sample mill according to the following procedure:

Prepare the coal sample to −1.18mm + 600 μm (−No. 16 + No. 30) size range. Weigh a 50g ± 0.01 gram of the sample, place it in the Hardgrove machine and grind the sample for 60 revolutions. Factor D is determined as the fraction of the coal passing through the sieve of 74 μm corresponding to 200 mesh. HGI is calculated from D as follows:

$$HGI = 13 + 6.93D \tag{3-3}$$

This procedure only results in relative values because the sampling mill is calibrated using a reference coal. The HGI of the reference coal is defined as 100.

The Hardgrove Grindability Index Standard Reference Sample (HGI-SRS) is a sample of coal used to calibrate instruments that are designed to determine the ease with which coal can be pulverized. Standard Reference Sample (SRS) representing grindability indices of approximately 40, 60, 180, and 100 shall be used (ASTM D409). Perform the grindability tests for SRS. Establish a calibration chart and equation using the sum of the least square method. The SRS can be ordered from the U.S. Department of Energy, Federal Energy Technical Center, Solid Fuel and Material Handling Division, Pittsburgh, PA. Figure 3.4 shows the machine design of a hardgrove grindability equipment. Figure 3.5 is an example of the calibration chart.

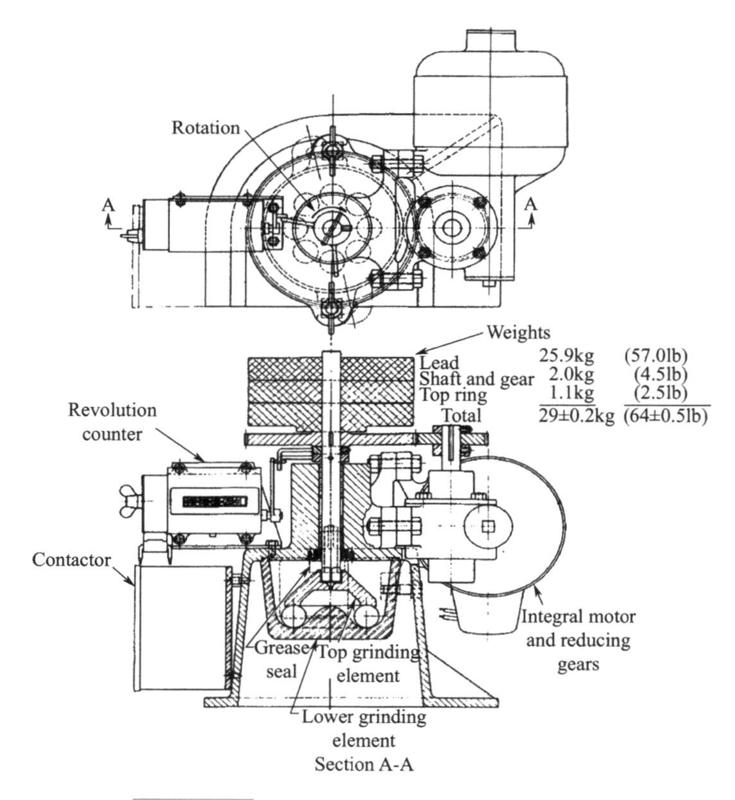

Figure 3.4 Hardgrove grindability machine

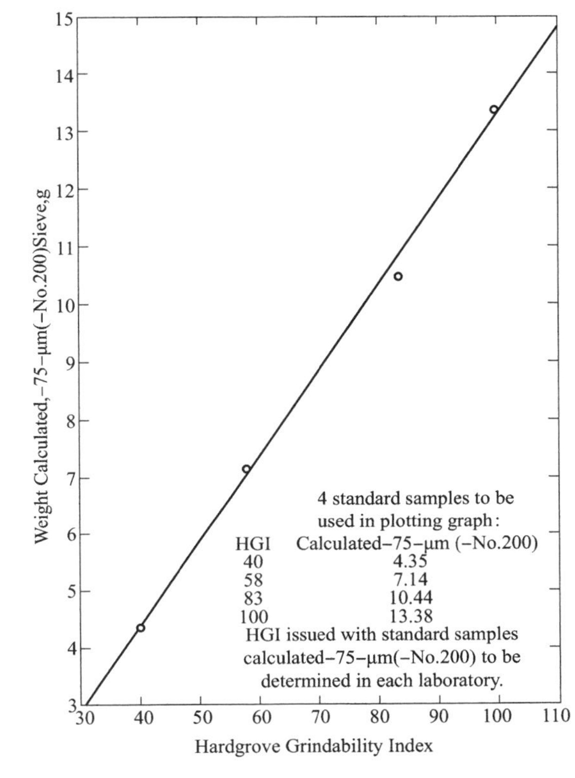

Figure 3.5 Example of A calibration chart

3.5 Mineral Matter in Coal and Ash Analysis

Coal also contains minerals, which mostly occur as inorganic crystallines and noncrystalline particles or masses. A coal seam may consist of as much as 50 percent minerals. At more than 50 percent mineral matter, the rock would be termed a carbonaceous shale rather than coal. Most mined coals are less than 20 percent mineral matter, and many coal contracts require less than 10 percent ash yield, which is approximately (but generally less than) the actual mineral matter content of a coal. More than 120 minerals and inorganic compounds have been documented in coals worldwide.

3.5.1 Mineral Matter in Coal

All coals contain noncombustible mineral matter. The residue from these minerals after coal has been burned is called ash. Ash differs in chemical composition and is usually less than the mineral matter originally present in coal. The amount of ash normally is determined by burning a 1~2 g sample of coal in a muffle furnace to the temperature of 750℃ as described in Section 3.3.1. During incineration, various weight changes take place, such as loss of water in the constitution of silicates, loss of carbon dioxide from the carbonates, oxidation of iron pyrites to iron oxide, and fixation of oxides of sulfur by bases such as calcium and magnesium. The Parr formula for calculating mineral matter is as follows (dry coal basis):

$$\text{Mineral Matter} = 1.08 \times \text{Ash} + 0.55 \times \text{Sulfur} \tag{3-4}$$

The mineral matters associated with coal consist of species belonging to shale, kaolin, sulfide, and chloride groups, as listed in Table 3.5. The ash is obtained by removing the organic coal substance by incineration at a low temperature. This technique is known as Low Temperature Ashing. This method consists of completely oxidizing the coal at 375℃ in a specially designed flask through which a controlled mixture of oxygen and nitrogen is passed. The total time required for complete oxidation is 2.5 hours.

Table 3.5 Typical minerals in coal

Minerals	Formula
Kaolinite	$Al_2O_3 \cdot 2SiO_2 \cdot H_2O$
Illite	$K_2O \cdot 3Al_2O_3 \cdot 6SiO_2 \cdot 2H_2O$
Muscovite	$K_2O \cdot 3Al_2O_3 \cdot 6SiO_2 \cdot 2H_2O$
Biotite	$K_2O \cdot MgO \cdot Al_2O_3 \cdot 3SiO_2 \cdot 2H_2O$
Orthoclase	$K_2O \cdot Al_2O_3 \cdot 6SiO_2$
Albite	$Na_2O \cdot Al_2O_3 \cdot 6SiO_2$
Calcite	$CaCO_3$
Dolomite	$CaCO_3 \cdot MgCO_3$
Siderite	$FeCO_3$
Pyrite	FeS_2
Gypsum	$CaSO_4 \cdot 2H_2O$
Quartz	SiO_2
Hematite	Fe_2O_3
Magnetite	Fe_3O_4
Rutile	TiO_2
Halite	NaCl
Sylvite	KCl

The composition of coal ash is determined by employing the Atomic Adsorption Spectrophotometer after digesting the coal ash using a wet chemical method. The typical ash compositions of high volatile bituminous coal, such as Pittsburgh seam coal, are shown in Table 3.6.

Table 3.6 Composition of high volatile bituminous coal ash

Ash Composition	Weight, %
Sulfur	10.87
SiO_2	3.53
Al_2O_3	37.64
TiO_2	20.11
Fe_2O_3	0.81
CaO	29.28
MgO	4.25
Na_2O	1.25
K_2O	0.80
Other	1.60
Total	95.74

3.5.2 Fusibility of Coal Ash

In preparing a sample for the coal ash fusibility test, first prepare 3 to 5 g of ash from thoroughly mixed air-dried coal, which has been ground to pass through a No. 60 sieve (250 μm). Spread the sample in a porcelain roasting dish. Place it in a coal muffle furnace and heat it gradually so that the temperature reaches 500℃ in 1 hour, and 750℃ in 2 hours. Ignite the ash to a constant weight at 750℃. Allow the ash to cool, transfer to an agate motor, and grind to pass a 150 μm sieve (U.S.A. series No. 100). Reignite the ash at 750℃ for 1 hour, cool rapidly, and immediately weigh portions for analysis. Thoroughly mix each sample before weighing.

Two solutions are prepared from the ash. Solution A is obtained by fusing the ash with sodium hydroxide (NaOH), followed by a final dissolution of the melt in dilute hydrochloric acid (HCl). Solution B is prepared by the decomposition of the ash with sulfuric acid (H_2SO_4), hydrofluoric acid (HF), and nitric acid (HNO_3). Solution A is used for the analysis of SiO_2 and Al_2O_3, and Solution B for the analysis of Fe_2O_3, TiO_2, P_2O_3, CaO, MgO, Na_2O, and K_2O. Atomic Adsorption Spectrophotometer can be used to determine those major elements in the ash sample. These ash compositions report to the fly ash and bottom ash in utility boilers, and report to the slag in the furnace, which then creates ash disposal, slag problems, de-rating, corrosive problems, and others.

The most fundamental measure of ash behavior with respect to "clinkering" tendency is the ash fusion temperature. Ash fusion is frequently used for the determination of the type of firing equipment and furnace design. To determine the ash fusion temperature, a sample of finely ground ash is pressed into a standard mold to form a slender vertical pyramid and set in an electrically heated furnace (Figure 3.6) (ASTM D1857). With a low rate of increasing temperature melting proceeds, four characteristic temperatures can be observed, depending on the shape of the deformed cone: (1) initial deformation temperature (IT), (2) softening temperature (ST)(H=W), (3) hemispherical temperature (HT)(H=1/2 W), and (4) fluid temperature (FT).

This test measures the temperature at which coal ash will soften and become fluid when heated under prescribed conditions. The ash fusibility is strongly dependent on the iron content of coal ash for both reduction and oxidation conditions. Table 3.7 lists typical ash fusion temperatures for Pittsburgh Seam Coal.

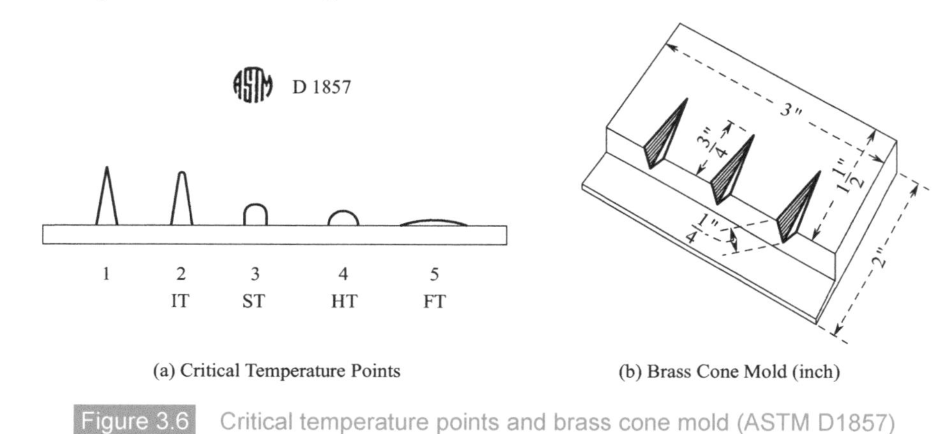

(a) Critical Temperature Points (b) Brass Cone Mold (inch)

Figure 3.6 Critical temperature points and brass cone mold (ASTM D1857)

Table 3.7 Ash fusion temperatures for pittsburgh seam coal, WV

Ash Fusibility	Temperature, °F	
	Reduction	Oxidation
Initial Deformation Temperature (IT)	2,030	2,265
Softening Temperature (ST)	2,175	2,385
Hemispherical Temperature (HT)	2,225	2,450
Fluid Temperature (FT)	2,370	2,540

3.6 Coal Ash Characterization

The properties of coal ash components are listed in Table 3.8.

Table 3.8 Properties of coal ash components

Element	Oxide	Melting Temp., °F	Chemical Property	Compound	Melting Temp., °F
Si	SiO_2	3120	acidic	Na_2SiO_3	1610
Al	Al_2O_3	3710	acidic	K_2SiO_3	1790
Ti	TiO_2	3340	acidic	$Al_2O_3 \cdot Na_2O \cdot 6SiO_2$	2010
Fe	Fe_2O_3	2850	basic	$Al_2O_3 \cdot K_2O \cdot 6SiO_2$	2100
Ca	CaO	4570	basic	Fe_2SiO_3	2090
Mg	MgO	5070	basic	$CaO \cdot Fe_2O_3$	2280
Na	Na_2O	sublimes at 2330	basic	$CaO \cdot MgO \cdot 2SiO_2$	2535
K	K_2O	decomposes at 660	basic	$CaSiO_3$	

3.6.1 Coal Ash Slagging Potential and Deposition Indicators

Slagging and fouling are the most common maintenance headaches in coal-fired boilers, which can negatively affect efficiency and production. Slag is molten ash and incombustible byproducts that remain following coal combustion. During operations, the ash experiences a range of high temperatures and depending on the ash properties, may result in troublesome deposits in different parts of the coal utilization or conversion equipment. The design of this equipment will involve the consideration of the melting or softening properties of the coal ash through the temperature range of the process. Since coal ash composition varies considerably, and the properties of the ash vary with composition, the melting properties also vary considerably.

The total weight of coal ash is calculated by

$$\text{Coal Ash} = SiO_2 + Al_2O_3 + Fe_2O_3 + CaO + MgO + Na_2O + K_2O + TiO_2 + P_2O_5 + SO_3 \tag{3-5}$$

Base to Acid Ratio is calculated by

$$\text{Base/Acid Ratio} = B/A = (Fe_2O_3 + CaO + MgO + Na_2O + K_2O)/(SiO_2 + Al_2O_2 + TiO_2) \tag{3-6}$$

Silica to Alumina Ratio is calculated by

$$\text{Silica/Alumina Ratio} = SiO_2 / Al_2O_3 \tag{3-7}$$

Iron to Calcium Ratio is calculated by

$$\text{Iron/Calcium Ratio} = Fe_2O_3/CaO \tag{3-8}$$

Iron to Calcium and Magnesium Ratio is calculated by

$$\text{Iron/Calcium magnesium Ratio} = Fe_2O_3/(CaO + MgO) \tag{3-9}$$

Dolomite percent is calculated by

$$\text{Dolomite percent (DP)} = (CaO + MgO)/(Fe_2O_3 + CaO + MgO + Na_2O + K_2O) \times 100\% \tag{3-10}$$

Equivalent Fe_2O_3 and Ferric percent (FP) is calculated by

$$\text{Equivalent } Fe_2O_3 = Fe_2O_3 + 1.11\ FeO + 1.43\ Fe \quad (3\text{-}11)$$

$$FP = Fe_2O_3/\text{Equivalent } Fe_2O_3 \times 100\% \quad (3\text{-}12)$$

Silica percent (SP) is calculated by

$$SP = SiO_2/(SiO_2 + \text{Equivalent } Fe_2O_3 + CaO + MgO) \times 100\% \quad (3\text{-}13)$$

Total Alkaline is calculated by

$$\text{Total Alkaline} = Na_2O + K_2O \quad (3\text{-}14)$$

3.6.2 Slagging Potential and Fouling Tendency of a Coal

Based on the different mechanisms involved in ash deposition on the heated surface, two general types of ash deposition have been defined as slagging and fouling. Slagging is the formation of molten or partially fused deposits on the furnace walls of a boiler or convection surfaces exposed to radiant heat. Fouling is defined as the formation of deposits on convection heat surfaces such as superheaters and reheaters.

Slags are the fused deposits or resolidified molten materials that form primarily on furnace walls or other surfaces exposed predominantly to radiant heat or excessively high gas temperature. Deposits are bonded, sintered, or cemented ash build-up that forms primarily on convective surfaces, such as superheater and reheater tubes, but also on furnace walls at lower than slag-producing temperatures.

Table 3.9 shows the criteria for the potential of slagging and fouling by using the Slagging Index and Fouling Index, which are related to the key mineral contents in the coal ash. The potential fouling is expressed by the concentration of Cl and Na_2O in the coal ash.

Table 3.9 Criteria for slagging and fouling potential Ash type

$Fe_2O_3 > CaO+MgO$	$Fe_2O_3 < CaO+MgO$
Slagging Index	
$R_s = \frac{Base}{Acid} \times S$	$R_s = \frac{(HT)+4(IT)}{5}$
$= \frac{Fe_2O_3 + MgO + Na_2O + K_2O + CaO}{SiO_2 + Al_2O_3 + TiO_2} \times S$	HT=Maximum hemispherical softening temperature
	IT=Minimum initial deformation temperature
$R_s < 2.0$—Low to medium slagging potential	
$2.0 \leqslant R_s \leqslant 2.6$—High slagging potential	$R_s > 2450$—Low slagging potential
$R_s > 2.6$—Severe slagging potential	$2450 \geqslant R_s \geqslant 2250$—Medium slagging potential
	$2250 > R_s \geqslant 2100$—High slagging potential

$Fe_2O_3 > CaO+MgO$	$Fe_2O_3 < CaO+MgO$
	$2100 < R_s$—Severe slagging potential
Fouling Index:	
$Rf=[Base/Acid] \times Na_2O$	When:
	$CaO+MgO+Fe_2O_3 > 20\%$ by weight of ash
$Rf < 0.5$—Low to medium fouling potential	Then:
$0.5 \leqslant Rf \leqslant 1.0$—High fouling potential	$Na_2O < 3$—Low to medium fouling potential
$Rf > 1.0$—Severe fouling potential	$3 \leqslant Na_2O \leqslant 6$—High fouling potential
	$Na_2O > 6$—Severe fouling potential
	When:
	$CaO+MgO+Fe_2O_3 < 20\%$ by weight of ash
	$Base/Acid < 0.5$ and $S < 1.0\%$
	Then:
	$Na_2O < 1.2$—Low to medium fouling potential
	$1.2 \leqslant Na_2O \leqslant 3$—High fouling potential
	$Na_2O > 3$—Severe fouling potential
Na_2O in Ash:	
< 1.0—Low to medium fouling potential	< 5.0—Low to medium fouling potential
1.0~2.5—High fouling potential	> 5.0—Severe fouling potential
> 2.5—Severe fouling potential	
Cl in Coal (independent of ash type):	
< 0.2 —Low	
0.2~0.3 —Medium	
0.3~0.5 —High	
> 0.5 —Severe	

(1) Types of ash deposits

① Fused slag deposits are formed on furnace walls and other surfaces exposed to predominantly radiant heat transfer. Slag deposits are usually associated with the physical transport of molten or tacky particles by the flue gases.

② High-temperature bonded deposits occur on convection heating surfaces, especially superheaters and reheaters. Alkali and calcium-bounded deposits are common in the United States.

③ Low-temperature deposits occur on air heaters and economizer. Condensation of acid or water vapor occurs on a cooled surface. Hard phosphatic deposits are extremely hard.

(2) Slagging potential of pyrite

Pyrite (FeS_2) in coal will oxidize in the high-temperature boiler to form molten Hematite as shown in Figure 3.7. The Molten Hematite forms small spherical droplets inside the boiler. These droplets have low drag and high specific gravity, and traverse across the gas flow stream to the boiler wall. The results of the droplets act as "glue" that collects other ash minerals and holds on to the "slag deposit".

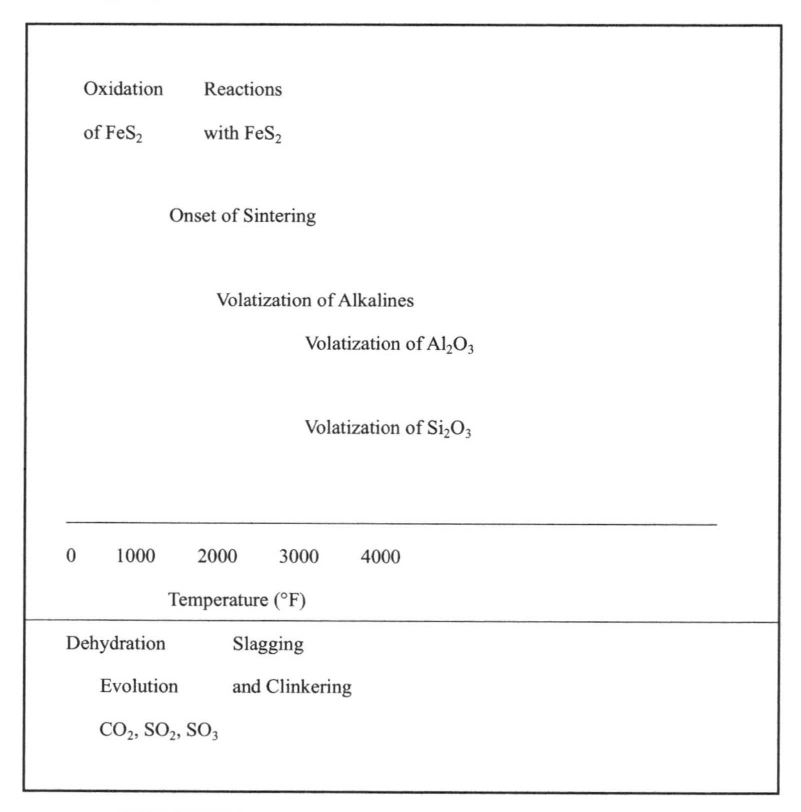

Figure 3.7 Effect of heating on coal mineral matter

$$FeS_2 + O_2 = SO_2\uparrow + Fe_2O_3$$
$$FeCO_3 + O_2 = CO_2\uparrow + Fe_2O_3$$

3.7 Characterization of Coal for Coke

Coke is a critical raw material to ironmaking blast furnace operation. It provides carbon sources for reducing iron ore and maintains the burden stability and permeability. With an increasing trend to use coal as alternative carbon sources by tuyere injection, the control of mechanical strength of coke is becoming more important for blast furnace operations.

3.7.1 Carbonization of Coal for Coke

If a coal is to be used successfully in the production of metallurgical coke, it must have the following attributes:

① Coking ability to yield a coherent coke alone or in blends with other coals.

② No expansion property to damage or cause an operating problem during carbonization.

③ Low sulfur and ash.

To determine the degree of consistency, sampling programs are instituted, and tests are devised to assay the quality of coal, including.

① Proximate Analysis.

② Ultimate Analysis.

③ Coal-Ash Composition.

④ Bulk Density.

⑤ Free Swelling Index.

⑥ Plastic Properties by Gieseler Plastometer.

⑦ Carbonization Testing - Sole-Heated Oven.

⑧ Swelling Properties of Coal using Dilatometer.

⑨ Petrographic (Microscopic) entities.

There are four qualities of coal that are of prime importance. These are:

① Coal volatility, affecting coke yield.

② Impurities in coal, affecting coke quality.

③ Composition of coal, affecting coke strength.

④ Characteristics of coal, affecting coke oven safety.

3.7.2 Fluidity or Plastic Properties of Coal by Constant-Torque Gieseler Plastometer (Audibert-Arnu Dilatometer)

This test is used to measure the plastic properties of coals that are related to the viscosity of the fluid as it is heated to the semicoke stage, about 500 ℃. This fluid state of coal during coking or pyrolysis is a definite characteristic of a coking coal.

Figure 3.8 shows the apparatus for tests in determining the coal fluidity (ASTM D2639). Prepare a coal sample according to ASTM D2639 to minus 425 μm (minus No. 40). Charge 5.0g of the coal sample into the crucible in the Gieseler Plastometer. Lower the crucible with the stirrer connected to a motor (a retort assembly) into a molten solid bath at 300 ℃. Heat the retort at a rate of (3.0 ± 0.1) ℃ /min on an overall basis. When the movement of the circular drum dial reaches 1.0 dial divisions per min, record the corresponding temperature as the "initial; softening temperature." Take the reading of temperature and counter dial movement at 1-minute intervals. Continue readings until the dial shows no further movement.

Record the maximum fluid temperature at which the dial movement reaches the maximum rate. This is the solidification temperature at which the dial movement stops. Maximum fluidity is the maximum rate of dial movement in dial divisions per minute (DDM).

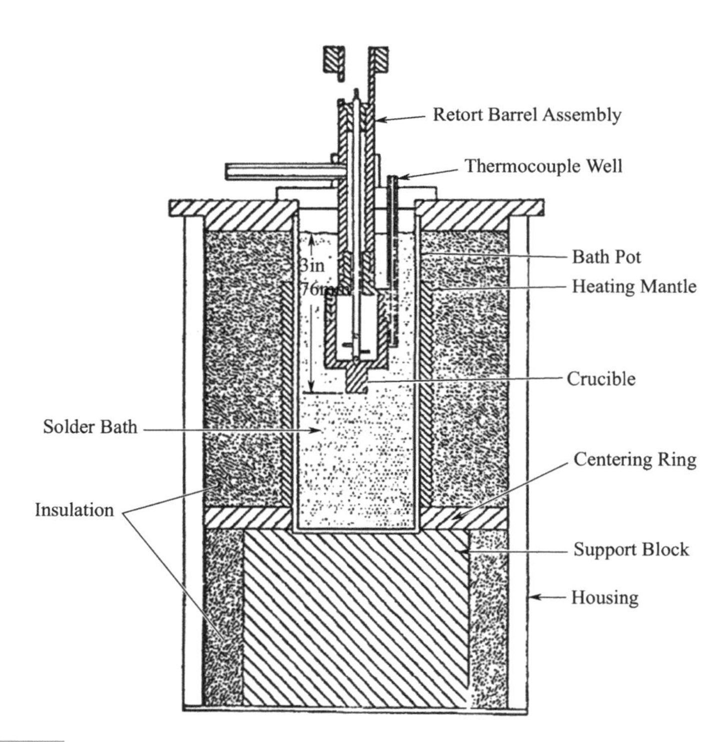

Figure 3.8 Furnace assembly for coal's fluidity test (ASTM Standard D2639)

3.7.3 Expansion and Contraction of Coal by the Sole-Heated Oven

A sole-heated oven, as depicted in Figure 3.9, is used to determine coal expansion during coking. Prepare a coal sample of 70%~85% passing 3.35mm (US No. 6) with a moisture content of 2%~4%, and charge ±0.05lb (0.023kg) coal sample into the carbonization chamber. Maintain the top of the sole-heated oven's temperature at 554 ℃ as the starting condition. Lower the piston onto the surface of the coal charge and place the total load over the plate (2.20 psi, or 1.52 kPa). The test is completed when the thermocouple on the top of the coal indicates 500 ℃. Measure the position of the piston at this moment and consider it as the definite datum for the calculation of total expansion or contraction. Record the time interval to this point.

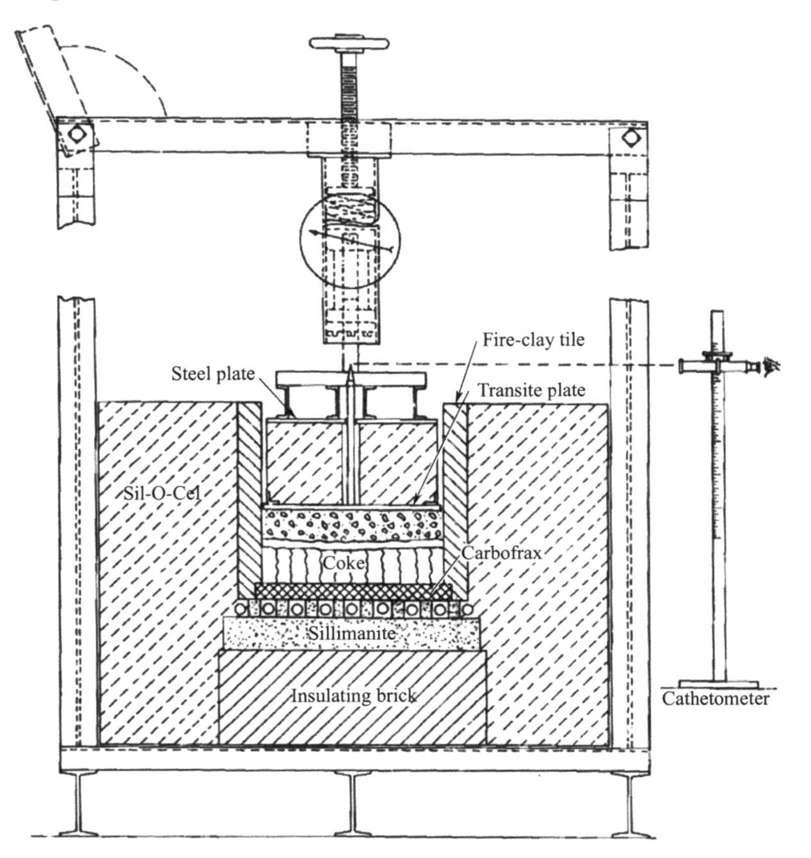

Figure 3.9 Sole-heated oven assembly for coal's expansion/contraction test (ASTM D 2014—96a)

The percentage expansion or contraction, E_t at the conditions of the test is calculated by the following equation at the moisture content of M_t % and bulk density of BD_t lb/ft^3:

$$E_t = 100\ (h_f - h_i)\ h_i \tag{3-15}$$

where h_f is the initial thickness of the coal charged in inches; and h_i is the final thickness of the coke in inches.

A positive value of E_t denotes an expansion and a negative value denotes a contraction. Figure 3.10 is an example of an expansion chart of different coal blends.

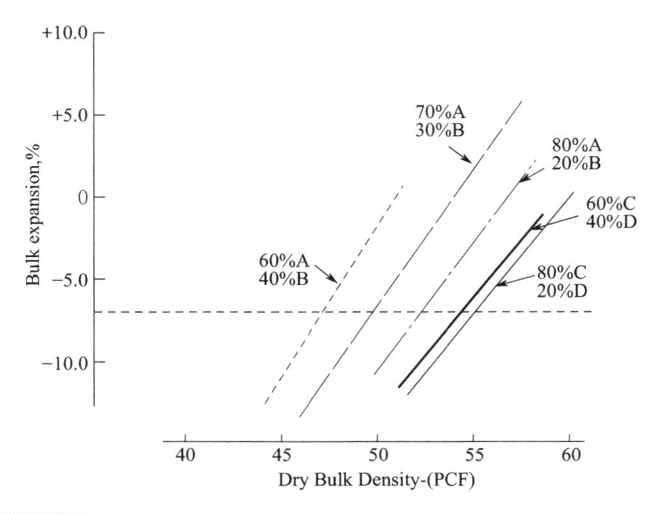

Figure 3.10 Example of coal expansion characteristics during coking

3.7.4 Swelling Properties of Coal Using a Dilatometer

The dilatation properties of coals may be used to predict or explain the behavior of a coal or blends during carbonization or in other processes such as gasification, liquefaction, and combustion.

The principle of the test method using a dilatometer is that the final volume of char obtained at the conclusion of a standard dilatation test is dependent on the mass of coal in the coal pencil and on the radius of the retort tube. This test method incorporates a procedure that determines the mass of air-dried coal in the coal pencil, provides a means to measure the average retort tube radius, and employs a means to report coal expansion on an air-dried coal weight basis.

The coal sample is pulverized to pass the No.60 (−250 μ m) US sieve. Coal pencil is made of about 10 grams of pulverized coal sample mixed with water (≤ 11% moisture) and formed by compression in a mold. The coal pencil is a cone and tapered shape, which is trimmed to 60mm long and placed into a retort tube. A steel rod piston is placed into a coal pencil retort tube, which is attached to a mechanical recording or transducer assemblies to record the movement of the piston and temperature. Load both retort tube and piston assemblies into the dilatometer furnace. The temperature of the retort furnace is established at 315℃ when the retort tube is lowered into the furnace. The temperature programmer is ramped at 3℃ per minute as soon as the sample is loaded into the retort furnace. The dilation curves will be recorded and provide the following information regarding the temperature, contraction, and dilation of coal:

Softening temperature T1 – The temperature at which the height of the coal pencil contracts 1.0% (0.06mm) from the highest recorded initial pencil height. In the instance of no dilation, terminate the test when the furnace temperature reaches 500℃.

Maximum contraction temperature T2 – The temperature at which the coal pencil starts to swell. Terminate the test when no movement of the piston can be detected for 5 minutes after the completion of the dilation.

Maximum dilation temperature T3 – The temperature at which the coal pencil first reaches the maximum height after swelling.

Percent contraction %C – The minimum recorded height of the char expressed as a percentage based on an initial coal pencil height of 60mm.

Percent dilation %D – The maximum recorded height of the char expressed as a percentage based on an initial pencil height of 60mm.

3.7.5 Free-Swelling Index (FSI)

The Free-Swelling Index (FSI) is used to obtain an indication of the free-burning or coking characteristics of coal when burned as fuel. It is also used to obtain an indication of the relative degree of oxidation of coal during storage. In general, FSI decreases, and the coking strength of coal diminishes as oxidation proceeds.

A translucent silica crucible and a silica ring-handle lid have an internal diameter of 11mm, a height of 26mm and a capacity of $17cm^3$. A sight tube with a dimension of 38mm (1.5″) ID×256mm (10″) length is needed for viewing the coke buttons. Any rigid and opaque material for the tube is preferred. The setup for the sight tube is shown in Figure 3.11.

Measure 1 gram of coal sample in the crucible and level by lightly tapping the crucible twelve times on the solid surface, rotating it between taps. Cover the crucible with a solid lid. Start the furnace at (820 ± 5)℃ for 1-1/2 minutes. Place the crucible in the furnace and set the timer for 2-1/2 minutes. Remove the coke button carefully. Make three buttons. View each coke button through the sight tube and compare it with the series of standard profiles shown in Figure 3.12 to determine the appropriate FSI.

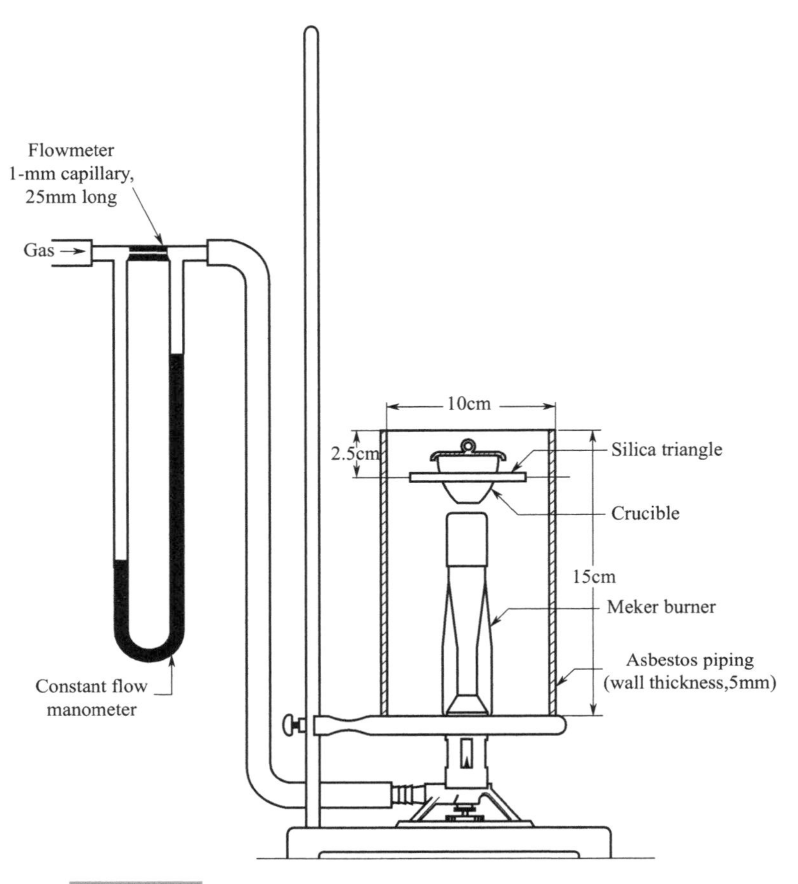

Figure 3.11 Burner assembly (ASTM Method D 2013—68)

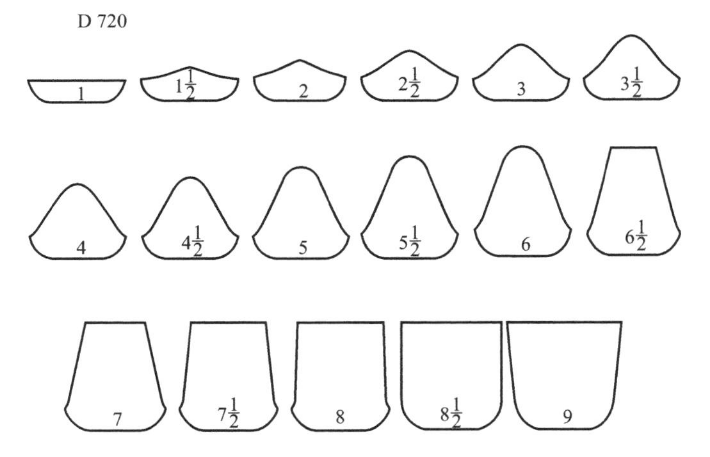

Figure 3.12 Full-scale standard profiles and corresponding FSI numbers (ASTM Method D 2013—68)

3.7.6 Stability Index for Coke Using the Tumbler Test

Relative measure of the resistance of coke to degradation by impact and abrasion is determined by the tumbler test. A 22-lb of coke is placed in a tumbler rotated at 14 RPM for 1400 revolutions. After tumbling, the coke product is sieved at 1-inch sieve. The percent of coke retained on the 1-inch sieve is taken as Stability Index. A coke with 50 percent or more of Stability Index is desired.

3.7.7 Metallurgical Coal

(1) Coke-making

Metallurgical coals are the type carbonized to make coke. The coke then becomes one of the three essential ingredients of the "charge" in the blast furnace at steel plants. These ingredients are coke, iron- bearing materials, and limestone or sand. Without metallurgical coals, steel cannot be made economically in commercial quantity in today's state of the art.

In the blast furnace, coke serves three functions that have a great deal to do in establishing specifications for the metallurgical coal that goes into coke-making. The coke provides:

① heat through combustion.

② a reducing agent of CO from carbon.

③ physical structure supporting the charge from the strong and porous material.

The coke-making process involves heating pulverized coal in the absence of air for a period of 14-19 hours in long narrow chambers called coke ovens. In the coking process, coal first softens and melts while simultaneously decomposing and resolidifying into porous but hard solid known as coke. When all the coal has been transformed into coke, it is pushed out from one end of the oven.

There are four qualities of coal that must be considered in order to make good coke:

① coal volatility, affecting coke yield.

② impurities in coal, affecting coke quality.

③ composition of coal, affecting coke strength.

④ characteristics of coal, affecting coke oven safety.

Unfortunately, there are very few metallurgical coals that have four qualities in the ideal composition. The usual practice is to blend several coals into a satisfying mixture based on an understanding of fundamental differences in compositions among the different types of coking coals.

(2) Preparation of an ideal coke oven charge to cover all the quality and safety factors

1) Coke yield

It is desirable to achieve the highest yield of coke practical per ton of coal charged to the coke oven. To produce a high coke yield, metallurgical coals should have a combination of characteristics that are reciprocal of one another. They should have a high carbon content and a low proportion of volatile matter, most of which go off as gases during the coking process.

Coke yield varies inversely with the volatile matter content of coal charge and can be calculated by

Coke yield, % = [100 − (volatile matter × 0.87)] × 100%

For coals with 17% and 35% volatile matter,

Coke yield, % = [100 − (17 × 0.87)] × 100% = 85%

Coke yield, % = [100 − (35 × 0.87)] × 100% = 70%

2) Composition of coal affecting coke strength

Strong abrasion coke is particularly necessary in a large blast furnace. The standard of strength is measured by the stability index. The higher the stability index, the better, and a good blast furnace coke should have a stability index of 50 or more. A "blocky" form of coke is desirable, but a relatively uniform size is important. Coke from high volatile matter coal has a low stability index, as low as 6, indicating weak coke. On the other hand, low and medium volatile matter coal has a stability index of 30-60 and their blends can assure the tumbler with a stability index of 50 or more.

Coal is made up of different organic substances transformed from vegetable debris by heat and pressure. Using a microscope, an experienced petrographer can readily identify and measure the amount of different entities or macerals, known as vitrinite, fusinite, exinite, and macrinite among others. Macerals have reactives-the binders like cement and water in concrete, and inerts- the aggregate like sand and gravel. The proper blend of reactives and inerts is needed to make a high-strength coke.

High volatile coals have an excess of reactives and a deficiency of inerts. Consequently, the stability index of coke made from it is low, but the range of stability index of 6-45 indicates the wide range of compositions among high volatile coals.

Low volatile coals are deficient in reactives but have an excess of inerts. Hence the stability index of the coke made from it is about 30-50.

Medium volatile coals tend compositionally toward either the high- or low-volatile type. But there are some coals such as Kopperston coal that have approximately the right proportion of reactives and inerts.

With the knowledge of the basic components of coals, coke makers can and do blend together in the right proportions, high, low, and medium volatile coals, to produce high strength cokes.

The fluidity − the characteristics of the coal to become soft and viscous as it is heated − may be an indicator of sorts such as how well different coals may blend together, but it is of limited value in predicting coke strength. The higher the dial divisions per minute (DDPM) number, the greater the fluidity. All of the strong coking coals are fluid enough to blend with most other coals.

The Free Swelling Index or "coke button" can quickly tell whether a coal is "coking" or not. The higher number is considered a stronger coke.

The reliable techniques for determining the strength of coke produced from metallurgical coals are coal petrography and pilot-scale coke ovens, which also are the best guides in evaluating metallurgical coals for the final crucial qualities that affect the safety of commercial coke ovens.

3) Coke oven safety

A commercial by-product coke oven battery represents an investment of many millions of dollars. Consequently, plant operators want to ensure that the metallurgical coals used will not damage the coke oven during coking. Coal may damage a coke oven during the coke-making process if the coal charged to the oven does not contract sufficiently. This is known as a "sticker". The repeated sticker can cause substantial damage. Whether or not the coke is a sticker, the coal charge may exert excessive pressure in the oven wall so that damage occurs during the coking process. High volatile coals exert a very low pressure, i.e., 1 psi or less. But low volatile coals exert a much higher pressure, 10 psi or more. A coal blend with a contraction range of 17%~12% is desirable.

The pressure developed by a coal blend cannot be calculated from the results of the individual coal. The coking pressure can only be reliably determined in a pilot-scale movable wall coke oven.

The metallurgical coal quality should be within the ranges listed in Table 3.10.

Table 3.10 Metallurgical coal quality

Quality Index	Range
Volatile matter	26%~30%
Sulfur	0.6%~0.9%
Ash	4%~7%
Expansion or contraction	−7%~12%
Free swell index	6~9
Fluidity	3,000~14,000

To obtain a high coke yield, it means low-volatile matter coals. To obtain a strong abrasive-resistant coke, it also means low volatile matter coals with right compositional combination of coals. However, low volatile matter coals are in short supply and usually higher priced. Therefore, enough medium volatile matter or high volatile matter coals with the relative low volatile matter is commonly used to make a satisfactory charge.

3.8 Coal Petrography

Coal is an organic rock. It consists of microscopically-distinct organic entities (petrological components) called macerals. Coal components can be identified microscopically by either of two techniques. The first is the thin-section or transmitted-light technique and the second is the polished-section or reflected-light technique. A maceral is a genetically-related group of carbonaceous entities with each entity also differing to varying degrees in chemical and physical properties.

3.8.1 Coal Petrographic Components

Coal is a mixture of macerals and minerals. The macerals (organic entities) are further subdivided into inert or reactive types, measured in terms of their appearances and reflectances in the polished section. The degree of "reactivity" is defined in reference to their susceptibility to physical or chemical changes under moderate heat. Table 3.11 lists four major lithotypes of coals.

Table 3.11 Lithotypes of coals

Lithotype	Minerals	Characteristics	Sp Gr and Ash Content
Vitrain	Vitrinite	Uniform shinny black bands	1.3 sp. gr., 0.5%~1% Ash
Clarain	Mainly vitrinite, with exinite	Laminated: composed of shinny and dull bands	About 1.3 sp. gr., 0.5%~2% Ash
Durain	Mainly inertinite, with exinite, little vitrinite	Dull, non-reflecting, poorly laminated	1.25~1.45 sp. gr., 1%~5% Ash mostly; 3%~5% Ash occasionally
Fusain	Fusinite	Charcoal-like fragments, soft fusain, hard fusain	1.35~1.45 sp. gr., 5%~10% or 1.60 and above, 55% or more Ash

(1) Four coal constituents

Vitrinite - macerals formed from wood plant tissues
Exinite - exines, made of resins, and waxes
Inertinite - components inert upon coking; macerals formed from fusinization
Mineral Matter - ash forming constituents

(2) Maceral grouping

① Vitrinite
Telinite - cell walls preserved
Collinite - homogeneous, structureless, most important
Vitrodetrinite - broken up, unidentifiable vitrinite

② Exinite - Darker maceral, lower carbon content, higher H_2 content
Sporinite - spore exines
Cutinite - outer covering of leaves, twigs, etc.
Resinite - resin
Alginite - algae
Liptodetrinite - broken up exinite macerals
③ Inertinite
Fusinite - charwood, burned woody tissue
Semifusinite - partially burned wood
Macrinite - humic "organic" mud, very fine grained
Sclerotinite - fungal sclerotia
Micrinite- very fine residue from devolatilization
Inerfoderinite - broken up inertinite macerals

(3) Mineral matter constituents

Clays, Quartz, Gypsum, Calcite, Hematite, Pyrite, FeS_2, Marcasite, FeS_2, etc.

3.8.2 Preparation of Coal Sample for Microscopical Analysis by Reflected Light- Stopes-Heerlen System

Materials needed for microscopic analysis are listed below:
Prepare coal sample to minus 850 μm (minus US No. 20).
Coal : Binder = 10g : 4g
Binder - Epoxy Resin (liquid)
Polishing Abrasives - aluminum oxide powders of 3, 1, 0.3, 0.1 μm sizes
Coal's components can be megascopically identified with the following features:
Banded Coal
Vitrain - uniform shinny layer
Attrital Coal - the ground mass or matrix of banded coal
Fusain - fibrous texture with a very dull luster
Non-banded Coal
Boghead - the waxy component
Impure coal includes:
Bone coal - impure coal that contains much clay or other fine grain detrital mineral matter
Mineralized coal - impure coal that is heavily impregnated with mineral matter interbedded Impurities

Mineral Parting - mineral breakage along specific crystallographic
planes in some specimens due to twinning,
exsolution lamellae, or chemical alteration.

Factors that Influence the Composition of West Virginia Coals

① Rank change across the state.

② Plant evolution.

③ Depositional environment.

The average composition of West Virginia coals is listed in Table 3.12.

Table 3.12 Average composition of West Virginia coals

Maceral Grouping	Weight, %
Vitrinite	72
Exinite	4
Inertinite	14
Minerals	10

3.8.3 Coal Types and Corresponding Coking Properties

Coal type is related to the type of plant material in the peat and the extent of its biochemical and chemical alteration. Its type can be assessed in terms of a variety of petrographic analyses. Coal petrology is concerned with the origin, composition, and properties of the distinct organic and inorganic components of different coals. To date, the principal practical application of coal petrology has been in the specification and selection of coals for carbonization.

Megascopically distinguishable ingredients of humic coals are recognized as vitrain, clarain, durain, and fusain. The four macroscopic components in coal are:

① Vitrain: Essentially bright glossy, brilliant in luster and homogeneous component of coal, having a massive texture and showing characteristic vitreous conchoidal fracture.

② Clarain: Bright component of coal in overall appearance, but less bright than vitrain. It is a heterogeneous material with a banded structure and has a definite and smooth surface when fractured at right angles to the bedding plane.

③ Durain: Essentially a dull component of coal, often with a suggestion of a slightly greasy black in overall appearance, and usually harder than bright coal. It is heterogeneous and has a firm granular texture.

④ Fusain: It occurs in pockets or as patches rather than uniform brand, of soft, somewhat fibrous material resembling charcoal. It is highly friable and can be readily powdered by fingers.

The sapropelic coals are divided into two groups: the "cannel coals" and the "boghead coals". The former is dullish black, with a slightly greasy appearance and conchoidal fracture; the latter is more brownish in color.

Just as a rock is composed of several minerals so is coal composed of several organic constituents termed macerals, the organic equivalent of minerals (which are different types of inorganic particles found in coals and other rocks).

The micro-components (macerals) found in high and medium rank coals are:

① Vitrinite (termed as huminite for peat and lignite or low rank coals, essentially woody materials): derived from plant cell substances varying in appearance from being completely structureless to exhibiting well discernible tissues. It is the major component of vitrain and one of the two principal components of clarain.

② Exinite (liptinite in low rank coals): derived from secretions and waxy coatings of plants, and lower in reflectance than vitrinite. It is the other principal component of clarain and durain.

③ Inertinite (derived mainly from oxidized plant material): with or without recognizable plant structures, and higher in reflectance than vitrinite. It is the major component of fusain and one of the two principal components of durain. In maceral analysis, it is commonly subdivided into maceral macrinite, micrinite, semifusinite and fusinite.

The principal techniques of applied coal petrology are listed below:

① Maceral analysis: It is a technique widely used for providing valuable information on the behavior of coals during carbonization. The analysis typically summarizes the total reactive (vitrinite + exinite + 1/3 semi-fusinite) and total inerts (2/3 semi-fusinite + other inertinite macerals + mineral matter).

② Microlithotype Analysis: Macerals rarely occur at random within any lithotype bands. Under the microscope, such a band can usually be seen to comprise thinner bands distinguished from each other of different maceral associations, known as microlithotype. This analysis is sometimes used as an additional aid in the study of coals for carbonization.

Determination of reflectance: The reflectance of each maceral varies directly with coal rank. Reflectance measurement is a valuable technique for determining the coal rank with precision. However, the macerals of any particular coal differ from each other in reflectance, and the determination of coal rank is, therefore, generally based upon vitrinite reflectance. In coal carbonization technology, reflectance distribution is frequently used to calculate the "strength index" (SI) and "composition balance index" (CBI), which are measures of the coking property of a coal or coal blend (Table 3.13).

In a given coal, exinite has a higher volatile matter than vitrinite, whereas inertinite has a lower value. The amount of mineral matter in most durains and fusains is considerably greater than that in the corresponding vitrains, while it is intermediate in clarains. Qualitatively, the mineral matter content in vitrain generally tends to be lower in silica and alumina and appreciably higher in alkalines than in durain, while clarain is again intermediate. The mineral matter in fusain varies widely in composition.

Table 3.13 Petrogical components of maceral and microlithotype

Principal Maceral groups	Principal Microlithotype groups
Vitrinite	Vitrinite
Vitrinite, exinite	Clarite
Vitrinite, inertinite	Vitrinertite
Vitrinite, exinite, inertinite	Trimacerite
Inertinite, exinite	Durite
Inertinite	Inertite

Lithotype	Principal microlithotype
Vitrite	
Vitrain	
Clarain	Vitrite
	Clarite
	Vitrinertite
	Trimacerite
	Durite
Durite	
Durain	
Fusain	trimacerite
	Inertite

The coking potentials of a coal can be predicted by the properties of the definable microscopic entities of which the coal is comprised. The entity composition of coal, in terms of the stages of coalification, can be determined by reflectance measurements. They are further separated into reactive and non-reactive classes on the basis of thermal properties (Table 3.14). The optimum inert content for each vitrinoid type is used to calculate a Composition Balance Index.

Table 3.14 Entity composition of a given coal

Class	Entity Composition	%	Summary
(Vitrinite)	Vitrinoid 9	10	
	Spouinite 10	40	
	Alginite 11	15	Total Reactive
Reactive			= 75.0%
	Exinoids	8	
	Resinoids .	1	
Semi-reactive	1/3-Semifusinoid	1	
	2/3-Semifusinoid	2	
Inerts	Micrinoids	15	Total Inert
	Fusinoids	5	= 25.0%
(Inertinite)	Mineral Matter	3	

COMMINUTION AND SIZE REDUCTION

4.1 Objective of Size Reduction

Size reduction is usually the first step in processing raw coal as it comes from the mine. Size reduction serves the following purposes:

① To facilitate cleaning of coal

② To liberate dissimilar minerals from the coal matrix for separation

③ To reduce to sizes that meet certain specifications

④ To facilitate handling such as to produce a particle size that will allow haulage by conveyor

During the comminution process, coal is reduced in size along its planes of weakness through pressure applied by a machine operating on various principles, including:

① Explosive disintegration

② Compression

③ Impact

④ Attrition (scrubbing between two hard surfaces)

⑤ Shearing or cutting

Figure 4.1 shows the basic concept of liberation of particle by size reduction: liberated coal, middlings (locked particles), and refuse (or tailings or impurity).

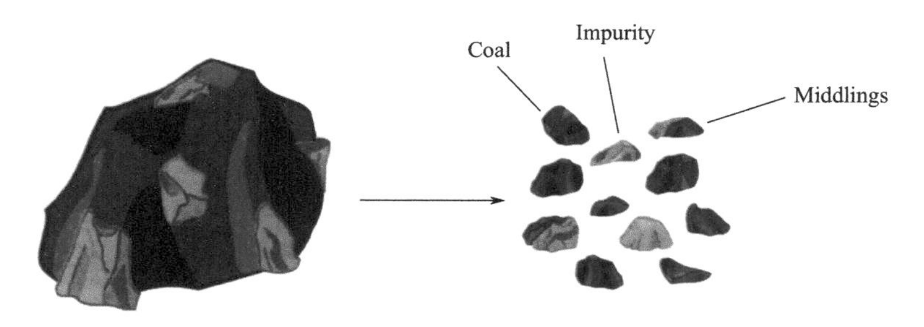

Figure 4.1 Liberation of coal particles

4.2 Energy Requirements for Size Reduction

Due to the heterogeneous nature of coal, it has been difficult to formulate generally applicable theories concerning the fundamentals of crushing and relate them to the variable factors involved. Hardness, as measured by the Hardgrove Grindability Index (HGI), is one parameter used to determine the resistance of coal to comminution. The energy requirement of size reduction can be estimated based on the concept of creating new surface areas during the size reduction. The criteria and theories of mechanical reduction including Rittinger, Kick and Bond have proved helpful in the estimation of energy requirements.

(1) Rittinger's criterion - The energy, E, required for the comminution is proportional to the new surface created

$$E \propto (A_2 - A_1) \tag{4-1}$$

$$E = k_1(A_2 - A_1) \tag{4-2}$$

where, A_1 and A_2 are the total surface areas of the particles before and after size reduction.

The specific surface area, S, is defined as

$$S = \frac{\text{total surface area of a given particle}}{\text{unit volume of a given particle}} = \frac{4\pi \frac{Dp^2}{2}}{\frac{4}{3}\pi Dp^3} = \frac{6}{Dp} \tag{4-3}$$

where, D_p is the diameter of the given particle.

The total volume of a given particle can be expressed as

$$V = \frac{m}{\rho_p} \tag{4-4}$$

where m is the mass and ρ_p is the density of the given particle. The total surface area of the given particle is given by

$$A = (\text{specific surface area}) \times (\text{total volume}) = \frac{6\lambda m}{\rho_p Dp} \tag{4-5}$$

where, λ is a shape factor.

Substitute Equation (4-5) into Equation (4-2) to yield

$$E = K_R\left(\frac{1}{Dp_2} - \frac{1}{Dp_1}\right) \tag{4-6}$$

(2) Kick's criterion - The energy required for size reduction is related to the reduction ratio.

$$E = K_K \ln\left|\frac{Dp_1}{Dp_2}\right| \tag{4-7}$$

where, D_{p1} and D_{p2} are the particle diameters of the feed and product. The particle diameter is determined by finding the screen size by which 80% of the material will pass through. The reduction ratio can be calculated by the following equations.

$$R = \text{Reduction Ratio} = \frac{\text{the maximum size of the feed}}{\text{the maximum size of product}} \tag{4-8a}$$

or

$$R_{80} = \text{Limited Reduction Ratio} = \frac{\text{the size of 80\% of the feed passing through}}{\text{the size of 80\% of the product passing through}} \tag{4-8b}$$

(3) Bond's criterion - The energy required to produce a particle of size D_p from a feed of infinite size is defined in equations (4-9) and (4-10).

$$E \propto \Delta\sqrt{Sp} \tag{4-9}$$

$$E = K_B\left(\frac{1}{\sqrt{Dp_2}} - \frac{1}{\sqrt{Dp_1}}\right) \tag{4-10}$$

Consider the general form of energy consumption during size reduction (Charles' criterion) as

$$\frac{dE}{dDp} = -KDp^n \quad \text{for } n = \alpha + 1\ (\alpha \geqslant 1) \tag{4-11}$$

① If n = −2 (Rittinger's criterion)

$$dE = -K\int_{Dp_1}^{Dp_2} Dp^{-2}dDp = K_R\left(\frac{1}{Dp_2} - \frac{1}{Dp_1}\right) \tag{4-12}$$

② If n = −1 (Kick's criterion)

$$dE = -K\int_{Dp_1}^{Dp_2} Dp^{-1}dDp = K_K \ln\left|\frac{Dp_1}{Dp_2}\right| \tag{4-13}$$

③ If n = −3/2 (Bond's criterion)

$$dE = -K\int_{Dp_1}^{Dp_2} Dp^{-3/2}dDp = K_B\left(\frac{1}{\sqrt{Dp_2}} - \frac{1}{\sqrt{Dp_1}}\right) \tag{4-14}$$

4.3 Principle of Size Reduction and Liberation

4.3.1 Reduction Ratio

Reduction ratio is defined as

$$\text{Reduction Ratio} = \frac{\text{"Top Size" of Feed}}{\text{"Top Size" of Product}} \tag{4-15}$$

① A particular type of machine is most efficient and performs best for a certain size range of feed.

② Size reduction is performed most efficiently in a series of stages; no machine can efficiently reduce the full range of size from the largest sizes to smallest sizes by repeated breakage.

③ A crusher design should be chosen to handle the hardest material in the coal but should not be over designed for what is needed.

4.3.2 Specific Energy

The specific energy of breakage of materials is defined as

$$\text{Specific Energy} = \frac{\text{Power Draw}}{\text{Product Rate}} \tag{4-16}$$

There are several theories, as derived in the previous section, relating the specific energy to feed and product "sizes" that have been conveniently reduced to the expression of

$$\frac{dE}{dDp} = -KDp^{n} \tag{4-17}$$

For Kick's criterion, n = −1, Equation (4-17) becomes

$$\Delta E_K = K_K(\ln Dp_1 - \ln Dp_2) \tag{4-18}$$

For Bond's criterion, n = −3/2, Equation (4-17) becomes

$$\Delta E_B = K_B\left(\frac{1}{\sqrt{Dp_2}} - \frac{1}{\sqrt{Dp_1}}\right) \tag{4-19}$$

For Rittinger's criterion, n = −2. Equation (4-17) becomes

$$\Delta E_R = K_R\left(\frac{1}{Dp_2} - \frac{1}{Dp_1}\right) \tag{4-20}$$

For Charles' criterion, n = $\alpha + 1$, Equation (4-17) becomes (for ores)

$$\Delta E_R = K_R\left(\frac{1}{Dp_2^{\alpha}} - \frac{1}{Dp_1^{\alpha}}\right) \tag{4-21}$$

where K_K, K_B, and K_R are constants.

Bond Work Index:

Bond Work Index is defined according to Bond's criterion in which n = −3/2 in Equation (4-17)

$$\Delta E_B = K_B\left(\frac{1}{\sqrt{Dp_2}} - \frac{1}{\sqrt{Dp_1}}\right) \tag{4-22}$$

$$W = 10 \times Wi\left(\frac{1}{\sqrt{(Dp2)_{80}}} - \frac{1}{\sqrt{(Dp1)_{80}}}\right) \tag{4-23}$$

$$P = T \times W \tag{4-24}$$

where $(Dp_1)_{80}$ = 80% passing size of feed, μm

$(Dp_2)_{80}$ = 80% passing size of product, μm

P = Power drawn, kW

T = Throughput of new feed, t/h

W = Work Input, kWh/t

Wi = Bond Work Index, kWh/t

The Bond Work Index values for some common materials are shown in Table 4.1.

Table 4.1 Some typical Bond Work Index values of different common materials

Material	Wi, kWh/t	with Correction factor	Wi, kWh/t
All Materials	13.8	Hardness Correlation	
Barite	6.2	Coal	11.4
Calcite	4.5	Coke	20.7
Cement Raw Material	10.6		
Cement Clinker	10.6	Softness Inversion	
Coal	11.4	Oil Shale	18.1
Coke	20.8	Shale	16.4
Copper Ore	13.1		
Dolomite	11.3	Shape Factor (Flat)	
Emery	58.2	Graphite	45.03
Galena Concentrate	3.2	Mica	13.45
Graphite	45.0		
Gravel	25.2	Cleavage	
Gypsum	8.2	Barite	6.24
Iron Ore -Taconite	14.9	Galena Concentrate	3.2
Limestone	11.6		
Mica	13.5	Shape Factor (round) (nip problem)	
Oil Shale	18.1	Gravel	25.2
Potash Ore	8.9	Silica Sand	16.5
Quartz	26.8	Compare Quartz	12.8
Shale	16.4		
Silicon Carbide	26.1		
Silica Sand	16.5		

An example of energy calculation for size reduction is given below:

Problem: 150 horsepower is required to handle the crushing of 1000 TPH (tons per hour) of ROM (run of mine) from an average size of 4-in to 1-in. Using the Rittinger's criterion (Equation 4-20) to determine the horsepower for the same crusher to take 1-in materials and reduce it to ¼-in at the same 1000 TPH rate.

Solution 1:

When Dp_1 = 4–in and Dp_2 = 1–in, and ΔE = 150 HP,

Specific Energy = 150 HP / 1000 ton = $\Delta E_R = K_R\left(\frac{1}{1}-\frac{1}{4}\right)$

Thus, K_R = 200/1000 = 1/5 HP-in/ton

Applying the Rittinger's criterion (Equation 4-20), $\Delta E_R = K_R\left(\frac{1}{D_{P_2}}-\frac{1}{D_{P_1}}\right)$

Specific energy = $\Delta E_R = \frac{1}{5}\left(\frac{1}{\frac{1}{4}}-\frac{1}{1}\right) = \frac{3}{5}$ HP / ton

Power draw = (3/5)(1000) = 600 HP

Energy is proportional to the increase in surface area, i.e., $\Delta E/\Delta A$ = Constant. This criterion can be used to deal with horsepower for a fixed period of time. E = HP × time.

Solution 2:

This problem can also be solved by logic giving through the process of reducing each dimension of a cube by ½. The final process evolves to this expression:

$$\frac{\Delta E_1}{\Delta A_1}=\frac{\Delta E_2}{\Delta A_2} \quad \frac{150}{288}=\frac{\Delta E_1}{1152} \quad \text{Therefore, } \Delta E=\frac{150\times 1152}{288}=600(\text{HP})$$

Total surface area of one 4-inch cube is

$(4'')^2$ (surface area) × 6 (sides of cube) × 1 (# of cube) = 96 in^2

Total surface area of 64 cubes for a 1-inch cube after crushing:

$(1'')^2 \times 6 \times 4^3$ (cube) = 484 in^2

$\Delta A_1 = 384 - 96 = 288\ in^2$

Total surface area of 4096 cubes for a 1/4-inch cube after crushing

$(1/4'')^2 \times 6 \times 16^3$ (cube) = 1536 in^2

$\Delta A_1 = 1536 - 384 = 1152\ in^2$

4.4 Unit Operations for Comminution

In this section, the principles of breakage and different comminution units that utilize those principles are discussed, including:

- Rotary Breaker;
- Single Roll Crushers;
- Double Roll Crushers;
- Jaw Crusher;
- Hammer Mills;
- Grinding Mills;
- Pulverizers;

- Attrition Mill.

Principles of Breakage:

Figure 4.2 shows the different comminution units that use impact, attrition, shear, and compression methods of material reduction. Each breakage method is described below.

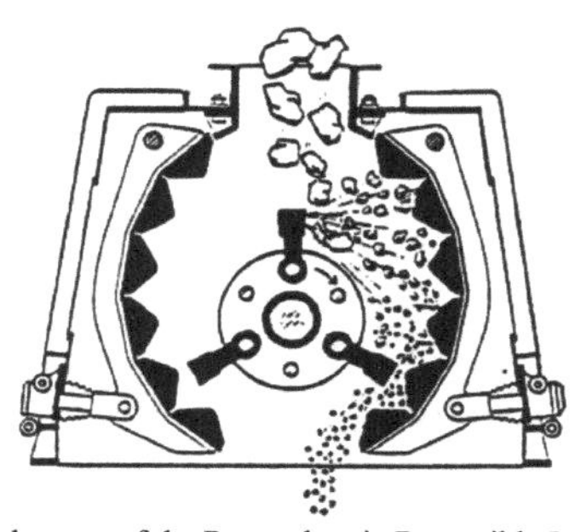

The bottom of the Pennsylvania Reversible Impactor is open and the sized material passes through almost instantaneously.Liberal clearance between hammers and the breaker blocks eliminates attrition,and crushing is by impact only.

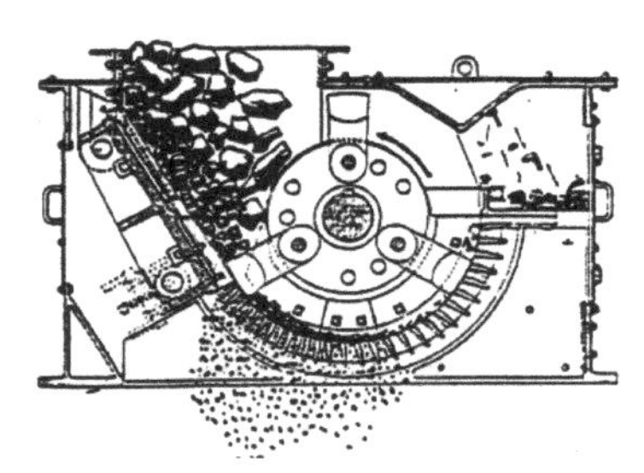

When a Pennsylvania Non-Reversible hammermill is used for reduction,material is broken first by impact between hammers and material and then by a scrubbing action(attrition)of material against screen bars.

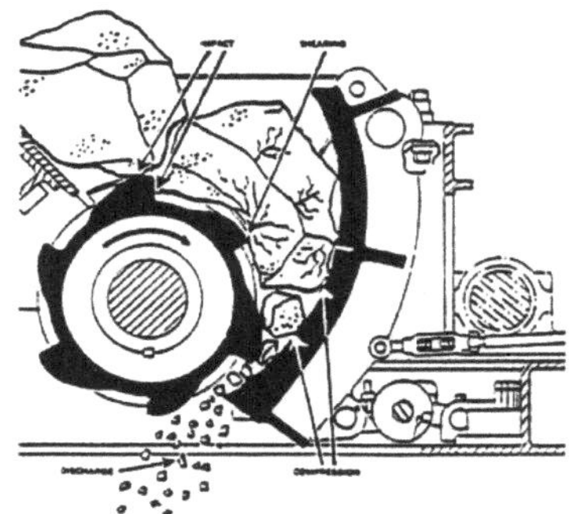

Reducing largefeed by a combination of shear,impact and compression.Hercules Single-Roll Crushers are noted for low headroom requirements and large capacity.

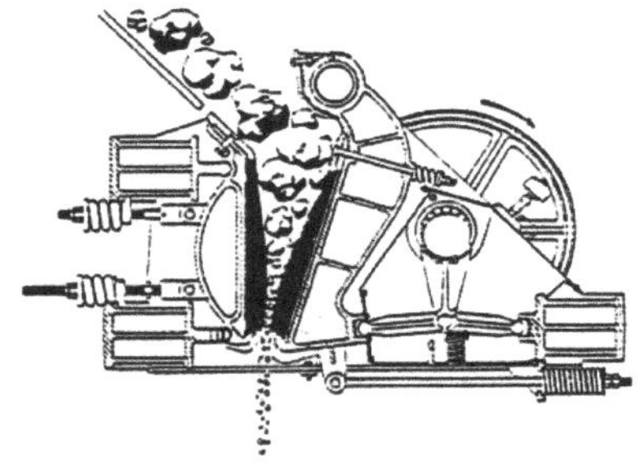

The Pennsylvania Jaw crushes by compression without rubbing. Hinged overhead and on the centerfine of the crushing zone,the swinging jaw meets the material firmly and squarely.There is no rubbing action to reduce capacity,to generate fines or to cause excessive wear of jaw plates.

Figure 4.2 Typical comminution units and related principles of breakage

(1) Impact

In crushing terminology, impact refers to the sharp, instantaneous impingement of one moving object against another. Both objects may be moving, such as a baseball bat hitting a fastball, or one object may be motionless, such as a rock being struck by hammer blows.

There are two variations of impact, including gravity impact and dynamic impact. Coal dropped onto a hard surface such as a steel plate is an example of gravity impact. Gravity impact is most often used when it is necessary to separate two materials with relatively different friability. The more friable material is broken, while the less friable material remains unbroken. Separation can then be achieved by screening.

The rotary breaker employs gravity impact only. This machine revolves so slowly that, for all practical purposes, gravity is the only accelerating force on the coal. Material dropping in front of a moving hammer (both objects in motion) illustrates the dynamic impact. When crushed by gravity impact, the free-falling material is momentarily stopped by the stationary object. But when crushing by dynamic impact, the material is unsupported, and the force of impact accelerates the movement of the reduced particles toward breaker blocks and/or other hammers.

Dynamic impact is the crushing method used by impactors. Dynamic impact has definite advantages for the reduction of many materials, and it is specified under the following conditions:

- When a cubical particle is needed.
- When finished products must be well graded and must meet intermediate sizing specifications as well as top and bottom specifications.
- When ores must be broken along natural cleavage lines in order to free and separate undesirable inclusions (such as mica in feldspars)
- When materials are too hard and abrasive for hammer mills, but where jaw crushers cannot be used because of particle shape requirements, high moisture content, or capacity.

(2) Attrition

Attrition is a term applied to the reduction of materials by scrubbing them between two hard surfaces. Hammer mills operate with close clearances between the hammers and the screen bars. They reduce particle sizes by attrition combined with shear and impact reduction. Though attrition consumes more power and exerts heavier wear on hammers and screen bars, it is practical for crushing less abrasive materials such as pure limestone and coal. Attrition crushing is most useful in the following circumstances:

- When a material is friable or not too abrasive.
- When a closed-circuit system is not desirable to control top size.
- When a maximum of fines is required.

(3) Shear

"Shear" consists of a trimming or cleaving action rather than the rubbing action associated with attrition. Shear is usually combined with other methods. For example, a single-roll crusher employs shear together with impact and compression. Shear crushing is normally called for under these conditions:

- When a material is somewhat friable and has a relatively low silica content for primary crushing with a reduction ratio of 4 to 1.
- When a minimum of fines is desired.
- When a relatively coarse product is desired (usually no finer than 1-1/2-in top size).

(4) Compression

As the name implies, crushing by compression is done between two surfaces, with the work being done by one or both surfaces. Jaw crushers using this method of compression are suitable for reducing extremely hard and abrasive rock. However, some jaw crushers employ attrition as well as compression and are not as suitable for abrasive rocks since the rubbing action accentuates the wear on crushing surfaces. As a mechanical reduction method, compression should be used in the following conditions:

- If the material is hard and tough.
- If the material is abrasive.
- If the material is not sticky.
- When a uniform product with a minimum of fines is desired.
- Where the finished product is to be relatively coarse. i.e., 1-1/2-in or larger top size.
- When the material is easy to break.

(5) Breaking and crushing

There are many types of breakers and crushers. The breakers are generally designed to reduce large lumps of a raw coal to a manageable size or to liberate the mineral matters for further processing, as are some crushers. Figure 4.3 shows the McNally rotary breaker. Figure 4.4 illustrates different types of impactors and crushers.

Figure 4.3 McNally rotary breaker

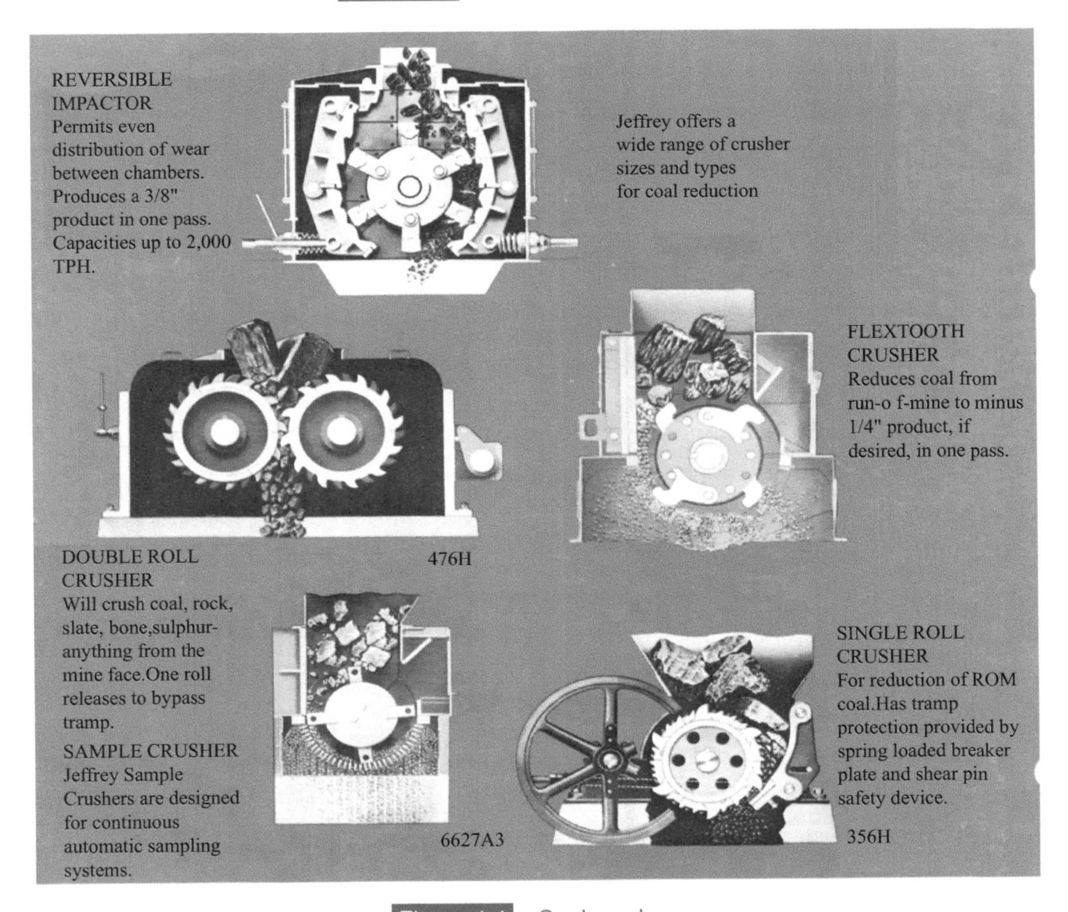

Figure 4.4 Coal crushers

A. Breakers

a. Rotary Breaker (Bradford Breaker)

The rotary breaker employs a slowly revolving cylinder that has a perforated plate for the passage of coal and a discharge chute for refuse. Lifters on the inside raise the coal and let it fall to the bottom, using gravity as the breaking force. Coarse refuse and tramp iron are discharged to a refuse chute. The advantages of rotary breakers are positive control of top size, breaking up large, laminated lumps separating the coal and rocks, and minimum maintenance on the breakers.

Generally, the rotary breakers are available in diameter from 9 ft to 14 ft and length from 12 ft to 28 ft. The correct length of each installation is determined by the tons-per-hour of feed, hardness and hardgrove grindability index of the coal, screen analysis of the run-of-mine coal, and the top size of the final product. The tonnage capacity can be from 400 to 2000 TPH. ASTM 440-86 for drop shattering test for sizing the breaker is typically used.

Utilizing the principle of a rotary breaker has the advantage of low fines production. Undersize is immediately screened out at the feed end. This prevents the small sizes from being ground down by large lumps and rocks. Gravity is the only breaking force used, which minimizes the production of fines.

b. Roll Type Breakers

The roll-type breakers employ double rolls, one of which is fixed and the other movable and adjustable over a large range. Relief springs on the movable roll absorb shocks from very large rocks.

Gearmatic Double Roll Breaker

This type of breaker has specially designed crushing segment teeth. Segmented teeth on each roll apply a splitting action to the lump of coal, breaking it apart freely without grinding off excessive fines. Undersized coal simply flows freely through the breaker along with the degradation. Screening ahead of this breaker is not necessary because the construction and operating characteristics allow the undersize to flow through the breaker itself without further degradation. Tonnages ranging from 750 to 1400 TPH can be handled in the breakers. The maximum lump size can be as large as 70 inches and the product size is 6 inches.

Double Roll Breaker

This heavy-duty roll breaker is designed to produce a high tonnage of crushed run-of-mine coal at a fast rate. It is designed to handle up to 2000 TPH of run-of-mine coal and reduce it to a nominal minus 8-inch without producing excessive fines.

B. Crushers

Crushing is defined as "the breakdown of particles into fragments with a top size greater than 1/20 of an inch (14 mesh Tyler)." The distance between the centers of the breaker rolls is about 30 inches. This has been increased to 34-in in the crushers so that small and shorter teeth could be used without changing the center to the center of the rolls.

Single Roll Crusher

It consists of a sturdy hopper with an internally mounted, renewable breaker plate as opposed to a frame-mounted crushing roll equipped with several long and a large number of short teeth across its full width. The coal is compressed between the revolving roll and the breaker plate. The long teeth act as feeders and penetrate the lumps, splitting them into smaller pieces. One design can accept run-of-mine feed. Others are specialized for re-crushing and may be adjusted to produce marketable sizes.

Double Roll Crusher

This type of crusher uses the impact of specially shaped teeth on both rolls to accomplish most size reductions. Compression action is secondary, thus minimizing the fines.

Triple Roll Crusher

Triple roll crushers can also accomplish two stages of reduction. Its single-roll crusher primary crushes the run of mine feed against a curved breaker plate. Double rolls mounted below perform a secondary reduction of a feed size as fine as 1/2-inch.

C. Mills

Hammer-Mill and Ring Crusher

These crushers accept mine feed from lump size downward. Both types use centrifugal force to deliver heavy blows to the feed while it is in suspension, driving the coal against a breaker plate until it is reduced to a size small enough to pass through the discharge. Grater bars may be included in the discharge opening to fix the maximum size. These types of crushers have a high capacity and produce greater fines than most other types of crushers. They need thorough maintenance and prompt replacement of parts subject to dulling by wear. The most popular use of them may be in crushing refuse to transportation sizes and preparing coal fines for the advanced coal froth flotation process.

4.5 Size Distribution Analysis

Accurate size data is essential in designing the flowsheet for a coal preparation plant. Plant designs seek to maximize the product yield while meeting a particular customer specification. Since washability studies are performed on all the size fractions of the plant feed except for ultrafine size fractions, the engineer needs accurate size data to gauge the cumulative effect of a particular unit operation on a range of size fractions.

In addition, all plant equipment is sized from the projected coal throughput of each of the three typical circuits (e.g., coarse, fine, and ultrafine). For example, when inaccurate size data leads the preparation engineer to underestimate the amount of coarse coal throughput, the plant's heavy-media bath will be overcrowded, resulting in an inefficient separation. This inefficiency may prevent the plant from meeting the customer's specifications and cause the cancellation of a contract. Conversely, if the preparation engineer overestimates the amount of coarse coal throughput, the fine and ultrafine circuits will probably be overloaded and cause a similar problem.

The most widely used measure for particle size analysis for coals is sieve sizing. The screens are available in 4 or 2 progressions of size. Particles are characterized as being greater than a given size and less than the next screen up, e.g., −1.0mm + 0.85mm (−No. 18 + No. 20 US sieve). The fraction-by-weight between sizes determined experimentally is accumulated and plotted as a cumulative distribution curve.

4.5.1 Sieve Analysis

Screening is one of the most common methods for particle size determination. To determine the particle size distribution, circular full or half-size sieves are available for use with a Rotap shaker. If a large amount of material (100 kg) coarser than 0.1 mm needs to be sized, Gilson screens may be used. For the materials with particle size less than 20 μm, micro-sieves can be employed. Wet sieving is necessary to obtain a large amount of materials of fine sizes.

Screen size is specified by either a linear dimension of square opening (aperture) or mesh. The mesh of the screen is the number of openings per linear inch. Various mesh designations are in use. A common feature is that the size of an opening can vary by a definite ratio when sieves are placed in a nested series. If the top screen in the nest is of size x_1, each sieve below decreases by a ratio of R such that the ith screen is of size $x_i = x_1(R)^{i-1}$. The value of R can be $1/\sqrt{2}$, $1/\sqrt[4]{2}$, $1/\sqrt[10]{10}$, or 1/2. For example:

When R is $1/\sqrt{2}$,

$$x_2 = x_1 \times \left(1/\sqrt{2}\right)^{2-1} = x_1 \times \left(1/\sqrt{2}\right)$$

$$x_1 = x_2 \times \sqrt{2}$$

$R = \dfrac{x_1}{x_2} = \sqrt{2} = 1.414$ for full size sieve

Use $R = \sqrt[4]{2} = 1.189$ for half size sieve

Figure 4.5 lists the ASTM E11 standard specification for sieve designation.

STANDARD SPECIFICATION FOR THE WOVEN WIRE TEST SIEVE CLOTH AND TEST SIEVES

ASTM E11

Nominal dimensions and permissible variations for sieve cloth (mesh) for Compliance, Inspection and Calibration grade test sieves

Sieve Designation		Nominal Sieve Opening	± Y Variation for Average Opening	+ X Maximum Variation for Opening	Resulting Maximum Individual Opening	Typical Wire Diameter	Permissible Average Wire Diameter	
U.S. Alternative	Standard						Min	Max
	Millimeters	inches	Millimeters	Millimeters	Millimeters	Millimeters		
5"	125	5.00	3.300	4.06	129.06	8.00	6.8	9.2
4.24"	106	4.24	2.800	3.59	109.59	6.30	5.4	7.2
4"	100	4.00	2.650	3.44	103.44	6.30	5.4	7.2
3-1/2"	90	3.50	2.390	3.18	93.18	6.30	5.4	7.2
3"	75	3.00	2.000	2.78	77.78	6.30	5.4	7.2
2-1/2"	63	2.50	1.690	2.44	65.44	5.60	4.8	6.4
2.12"	53	2.12	1.420	2.15	55.15	5.00	4.3	5.8
2"	50	2.00	1.340	2.06	52.06	5.00	4.3	5.8
1-3/4"	45	1.75	1.210	1.91	46.91	4.50	3.8	5.2
1-1/2"	37.5	1.50	1.010	1.67	39.17	4.50	3.8	5.2
1-1/4"	31.5	1.25	0.855	1.47	32.97	4.00	3.4	4.6
1.06"	26.5	1.06	0.722	1.29	27.79	3.55	3.0	4.1
1.00"	25	1.00	0.682	1.24	26.24	3.55	3.0	4.1
7/8"	22.4	0.875	0.613	1.14	23.54	3.55	3.0	4.1
3/4"	19	0.750	0.522	1.01	20.01	3.15	2.7	3.6
5/8"	16	0.625	0.441	0.89	16.89	3.15	2.7	3.6
0.530"	13.2	0.530	0.365	0.78	13.98	2.80	2.4	3.2
1/2"	12.5	0.500	0.346	0.75	13.25	2.50	2.1	2.9
7/16"	11.2	0.438	0.311	0.69	11.89	2.50	2.1	2.9
3/8"	9.5	0.375	0.265	0.61	10.11	2.24	1.9	2.6
5/16"	8	0.312	0.224	0.54	8.54	2.00	1.7	2.3
0.265"	6.7	0.265	0.189	0.48	7.18	1.80	1.5	2.1
1/4"	6.3	0.250	0.178	0.46	6.76	1.80	1.5	2.1
#3.5	5.6	0.223	0.159	0.42	6.02	1.60	1.3	1.9
#4	4.75	0.187	0.135	0.37	5.12	1.60	1.3	1.9
#5	4	0.157	0.114	0.33	4.33	1.40	1.2	1.7
#6	3.35	0.132	0.096	0.29	3.64	1.25	1.06	1.50
#7	2.8	0.110	0.081	0.26	3.06	1.12	0.95	1.30
#8	2.36	0.0937	0.069	0.23	2.59	1.00	0.85	1.15
#10	2	0.0787	0.059	0.20	2.20	0.90	0.77	1.04
#12	1.7	0.0661	0.050	0.18	1.88	0.80	0.68	0.92
#14	1.4	0.0555	0.042	0.16	1.56	0.71	0.60	0.82
#16	1.18	0.0469	0.036	0.140	1.320	0.63	0.54	0.72
#18	1	0.0394	0.030	0.130	1.130	0.560	0.480	0.64
	Micrometers	inches	Micrometers	Micrometers	Micrometers	Millimeters		
#20	850	0.0331	26.2	114	964	0.500	0.43	0.58
#25	710	0.0278	22.2	101	811	0.450	0.38	0.52
#30	600	0.0234	19.0	91	691	0.400	0.34	0.46
#35	500	0.0197	16.2	80	580	0.315	0.27	0.36
#40	425	0.0165	14.0	73	498	0.280	0.24	0.32
#45	355	0.0139	12.0	65	420	0.224	0.19	0.26
#50	300	0.0117	10.4	58	358	0.200	0.17	0.23
#60	250	0.0098	8.9	52	302	0.160	0.13	0.19
#70	212	0.0083	7.8	47	259	0.140	0.12	0.17
#80	180	0.0070	6.8	43	223	0.125	0.106	0.150
#100	150	0.0059	6.0	38	188	0.100	0.085	0.115
#120	125	0.0049	5.2	34	159	0.090	0.077	0.104
#140	106	0.0041	4.7	31	137	0.071	0.060	0.082
#170	90	0.0035	4.2	29	119	0.063	0.054	0.072
#200	75	0.0029	3.7	26	101	0.050	0.043	0.058
#230	63	0.0025	3.4	24	87	0.045	0.038	0.052
#270	53	0.0021	3.1	21	74	0.036	0.031	0.041
#325	45	0.0017	2.8	20	65	0.032	0.027	0.037
#400	38	0.0015	2.6	18	56	0.030	0.024	0.035
#450	32	0.0012	2.4	17	49	0.028	0.023	0.033
#500	25	0.0010	2.2	15	40	0.025	0.021	0.029
#635	20	0.0008	2.1	13	33	0.020	0.017	0.023

Figure 4.5 ASTM E11 Standard specification for the woven wire test sieve cloth and test sieves

4.5.2 Rosin-Rammler Equation

It is usual to use log-log, semi-log, or Rosin-Rammler scales to construct the size distribution curves because they produce flat curves. The Rosin-Rammler scale and Tyler scale are frequently used to present the size distribution of coal. Coal generally breaks into sizes that conform to an exponential relationship that was modeled by Rosin and Rammler in 1933 and further modified by Bennet as follows:

$$R(x) = e^{-(x/x_0)^n} \tag{4-25}$$

where $R(x)$ is the cumulative weight fraction of coal retained on a sieve of opening size x. This value increases as x decreases, as shown by the size analysis in Table 4.2.

Table 4.2 Size distribution data for a given raw coal

Size x, mm	U.S. Series Sieve No.	Tyler Series Mesh	Weight Retained , g	Weight Retained r(x)	Cumulative Weight Retained R(x)
25	1-in.		301.8	0.0951	0.0951
12.5	1/2-in		559.9	0.1764	0.2715
6.3	1/4-in.		751.4	0.2367	0.5082
0.6	30	28	1102.2	0.3472	0.8554
0.3	50	48	136.9	0.0431	0.8985
0.15	100	100	166.7	0.0525	0.951
0.075	200	200	64.8	0.0204	0.9714
−0.075	−200	−200	90.8	0.0286	1
Total	-	-	3174.4	1.00	-

Taking nature log against both sides of Equation (4-25) (the Rosin-Rammler-Bennet equation),

$$\ln R(x) = -(x/x_0)^n \tag{4-26}$$

Repeating this operation,

$$\ln\left(\ln\left(\frac{1}{R(x)}\right)\right) = n\ln\left(\frac{x}{x_o}\right) \tag{4-27}$$

Equation (4-27) is a linear function of the screen opening x, having a slope n. The coal preparation engineers use a specially pre-ruled paper, known as "Rosin-Rammler Paper", to obtain this straight-line graph representing a coal size. Every run-of-mine coal has a unique "fingerprint" characterized by the slope, n, and by x_o, i.e., the absolute size constant. Mathematically, x_o is the value of x when R(x)=0.3679. It occurs when the exponent of Equation (4-25) equals Equation (4-28).

$$R(x) = \frac{1}{\exp(1)} = \frac{1}{2.71828} = 0.3679 \tag{4-28}$$

For convenience, the Rosin-Rammler paper is printed with a slightly modified version of Equation (4-25), enabling the preparation engineers to work with weight percentages instead of weight fractions.

$$R(x)(\%)=e^{-(x/x_0)^n}\times 100\% \tag{4-29}$$

The particle size distribution of a coal sample is plotted as x vs. R(x), as shown in Figure 4.6. The absolute size constant is read from the screen opening from the graph at 36.78% of oversize percent, while the slope n is determined from the ratio of Δy and Δx, directly read out from the scale of the ruler used.

The slope of the line is n. x_0 is the value of x when R(x)=36.79%. It is the value when $(x/x_0)^n=1$. The slope, n, is a measure of the degree of dispersion of the particle size. It is readily seen that the limits for n are 0 and ∞. If $n\to\infty$, the curve will be plotted as a vertical line. All particles will approach the same size and there will be no size distribution.

If n = 1, the theory is simplified to the ideal one of breakage in which an absolute random distribution of forces in the broken material may be assumed. As $n\to 0$, n becomes very small, and the particle size will spread over a wide range. The size of every particle will differ relatively small from that of the next larger or smaller particle.

There is a tendency for the plot to depart from a straight line in large sizes. It is reported that n increases as particles become very fine, so that at the fine end, there may also be a departure from linearity. Thus, over a very wide size distribution, the Rosin-Rammler plot of an actual cumulative coal-particle size distribution on the Bennett diagram paper tends to be S-shaped.

Results are generated in Table 4.2 using sized experiment data and data analysis of weight retained on sieves. The particle size distribution curve can be established in the Rosin-Rammler-Bennet graph. For the determination of the slope of the curve, use any ruler to read the values of Δy and Δx in the unit of choice to obtain $n=\Delta y/\Delta x=1.9\ \text{cm}/2.6\ \text{cm}=0.731$. The value of x_o is determined by reading the particle size when R "oversize" wt.% at 36.78%. This resulted in x_o = 9 mm. Therefore, the Rosin-Rammler-Bennet equation is

$$R(x)(\%)=e^{-\left(\frac{x}{9.0}\right)^{0.731}}\times 100\% \tag{4-30}$$

The derived Rosin-Rammler-Bennet equation can be a useful tool to predict the weight of the given particle size range. For example, the weight of particle size range of −250 μm+105 μm (-US series no. 60+no.100 sieves) can be calculated by using Equation (4-26). R(x) represents the percent retained on the screen x. Therefore

For x = 0.250 mm: $R(0.25)\ \% = \exp[-(0.250/9)^{0.731}]\times 100\%$

$= \exp[-0.07284]\times 100\% = 92.98\%$

For x = 105 mm: $R(0.105)\ \% = \exp[-(0.105/9)^{0.731}]\times 100\%$

$= \exp[-0.03863]\times 100\% = 96.21\%$

The weight between −250 μm+105 μm (-US series no. 60+no.100) is

$R(0.105)\ \% - R(0.250)\ \% = 96.21\ \% - 92.98\ \% = 3.23\ \%$

The Rosin-Rammler-Bennet Equation can be linearized as:

$$\ln\left[\ln\left(\frac{1}{R(x)}\right)\right]=n\left[\ln\left(\frac{1}{x_0}\right)+\ln(x)\right] \tag{4-31}$$

$$Y=b+mX \tag{4-32}$$

Table 4.3 lists the size distribution for linear curve fitting based on equation (4-32).

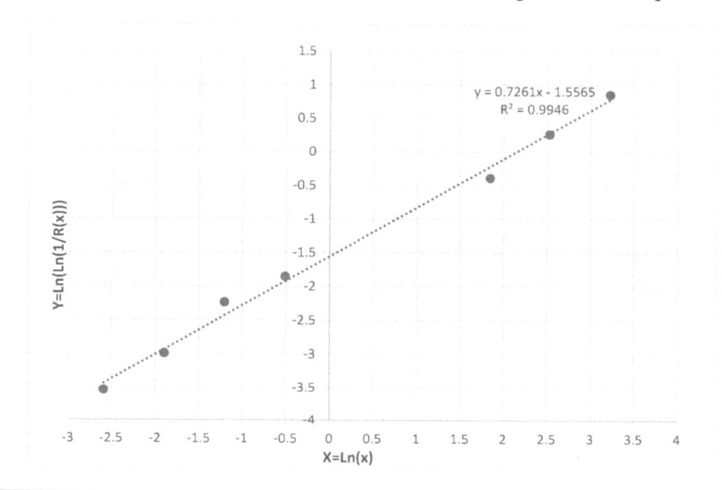

Figure 4.6 Curve fitting of Equation (4-32) based on data in Table 4.3

Table 4.3 Linearization of the Rosin-Rammler-Bennet size distribution for linear curve fitting

Size x, mm	X	R (x)	1/R(x)	ln(1/R(x))	Y
25	−2.590	0.0951	10.5182	1.022	−3.540
12.5	−1.897	0.2715	3.6839	0.566	−2.991
6.3	−1.204	0.5082	1.9679	0.294	−2.235
0.6	−0.511	0.8554	1.1691	0.068	−1.857
0.3	1.841	0.8985	1.113	0.046	−0.390
0.15	2.526	0.951	1.0515	0.022	0.265
0.075	3.219	0.9714	1.0295	0.013	0.856

SCREENING, CLASSIFICATION, AND DESLIMING

5.1 Screen Efficiency

There are two common methods of calculating screen efficiency, depending upon whether the desired product is the "oversize" or "undersize" (through) from the screen deck. If the oversize is considered to be the product, the screen operator wants to remove as much as possible of the undersize material. In that instance, the screen performance is expressed as "efficiency of undersize removal." When the undersize (through) is considered to be a product, the operator wants to recover as much of the undersize as possible. In this section, the case of "efficiency of undersize recovery" is used to illustrate the procedures.

The efficiency of undersize removal is determined by taking a sample of the oversize off a screen deck and making a sieve test to determine the screen analysis. The analysis will show if all the oversize products are the desired ones. In other words, it will show the content of undersized particles in the oversized product. No screen is commercially capable of removing all the undersize.

To improve efficiency, either the screen area must be increased, or the amount of feed must be decreased. That is because a screen operates at peak efficiency with a material load of approximately 80% of that at which the screen was calculated. As shown in Figure 5.1, at point "a", screening efficiency achieves a maximum value. It is a point where the screen operates at approximately 80% of the rated capacity. When sizing a new screen, it can be assumed that the customer does not want to reduce capacity or feed throughput. This leaves only the option of increasing the screening area as a means of improving screen efficiency. Therefore, when the oversize material in the feed is 30% or less, or when maximum screen efficiency is required, regardless of the amount of oversize materials in the feed, increasing the calculated screen area by 20% will result in improved efficiency.

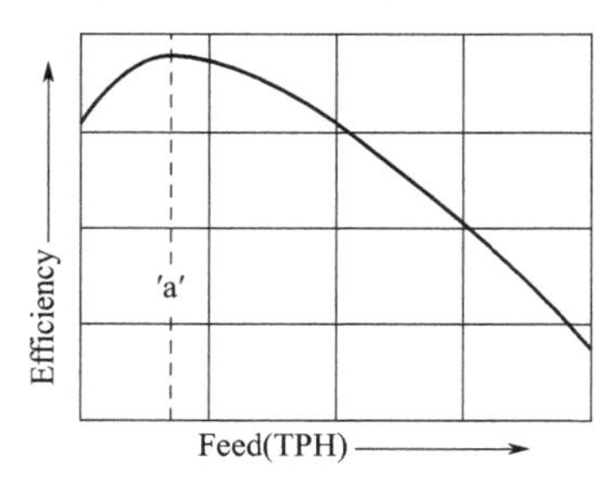

Figure 5.1 Screen efficiency vs feed capacity

5.1.1 Efficiency of Undersize Removal

The efficiency of undersize removal covers the application where undesired fines are to be removed from the oversize product. The definition can be expressed in the following equations:

$$E_m = 100\% - b\% \tag{5-1}$$

where, E_m = Efficiency of undersize removal, E_m= 100% − b%, b = percentage of undersize in oversize products

or,

$$E_m = \frac{\text{\% or TPH of oversize material in the feed}}{\text{\% or TPH of the feed retained as the oversize product}} \times 100\% \tag{5-2}$$

Example 1:

Given: The feed to the screen for 1-inch separation is 100 TPH. The sieve analysis for the feed is given in Table 5.1. And the product distribution from the screen analysis is shown in Figure 5.2.

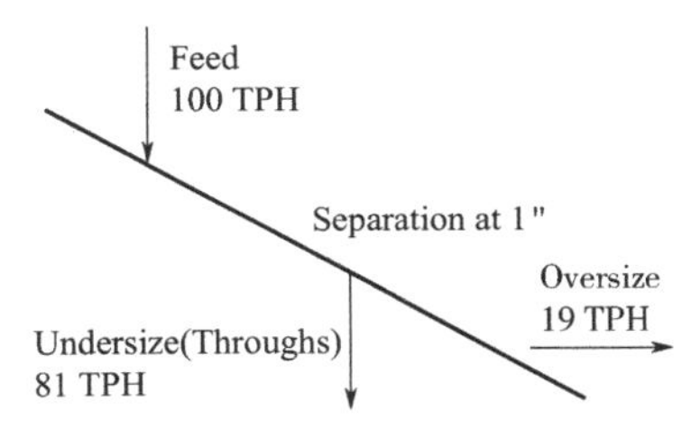

Figure 5.2 Diagram of screen separation example 1

Table 5.1 Sieve analysis of the feed

Size, inch	Percentage Passing
2	100
1	90
½	50
¼	20

The efficiency of undersize removal E_m is calculated from either Equation (5-1):

$$b\% = \frac{\text{misplaced material in the oversize product}}{\text{actual oversize product}} \times 100\% = \frac{90-81}{19} \times 100\% = 47\%$$

$$E_m = 100\% - b\% = 100\% - 47\% = 53\%$$

or Equation (5-2):

$$E_m = \frac{\text{theoretical oversize product}}{\text{actual oversize product}} = \frac{10}{19} \times 100\% = 53\%$$

Example 2:

Given: The feed to the screen for 1-inch separation is 100 TPH. The sieve analysis for the feed is given in Table 5.2. And the product distribution from the screen is shown in Figure 5.3.

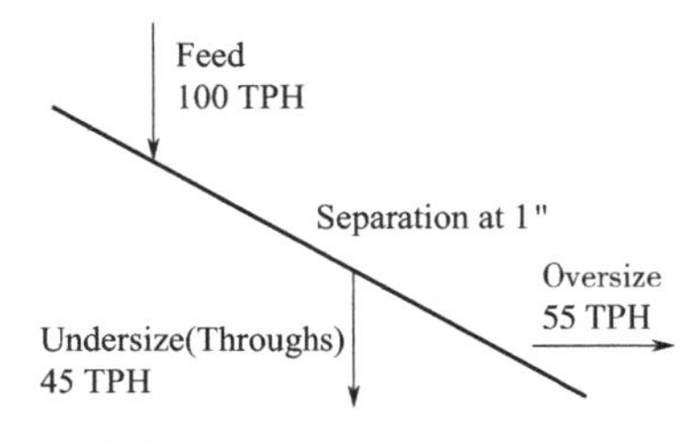

Figure 5.3 Screen Separation example 2

Table 5.2 Sieve analysis of the feed

Size, inch	Percentage Passing
2	100
1	50
½	30
¼	10

From Equation (5-1):

$$b\% = \frac{50-45}{55} \times 100\% = 9\%$$

$$E_m = 100\% - 9\% = 91\%$$

From Equation (5-2):

$$E_m = \frac{50}{55} \times 100\% = 91\%$$

When the amount of oversize is small, the resulting efficiency appears to be low, as shown in Example 1, E_m=53%. Where there is a large amount of oversize product, the resulting efficiency of undersize removal appears to be high, as shown in Example 2, E_m=91%. In both cases, the actual amount of undersize in the oversize product is relatively small, 9 TPH and 5 TPH, respectively. The apparent low efficiency in Example 1 can be improved, as previously discussed, by increasing the screening area by 20%.

5.1.2 Efficiency of Undersize Recovery

The efficiency of screen is linked to various factors, such as the vibration of the screen, the bed depth of material, the moisture of feed material, and so on. Across the screen deck, the function of different sections of the screen deck is not the same (Figure 5.4). When the feed material enters the screen, it starts stratification. Then, the screen starts saturation screening. In the remaining part of the screen, the undersize material is screened out via repetitive trials of screening.

The efficiency of undersize recovery is used for the cases where the undersize is the product. It is defined as

$$E_r = \frac{\text{Actual amount of screen undersize product}}{\text{Theoretical amount of undersize material in feed}} \times 100\% \quad (5\text{-}3)$$

or

$$E_r = \frac{(a-b)\times 100}{a\times(100-b)} \times 100\% \quad (5\text{-}4)$$

where

a = percentage of theoretical undersize in the feed as a percentage of feed.

b = percentage of actual undersize in oversize product as a percentage of oversize.

E_r = Efficiency of undersize recovery, %

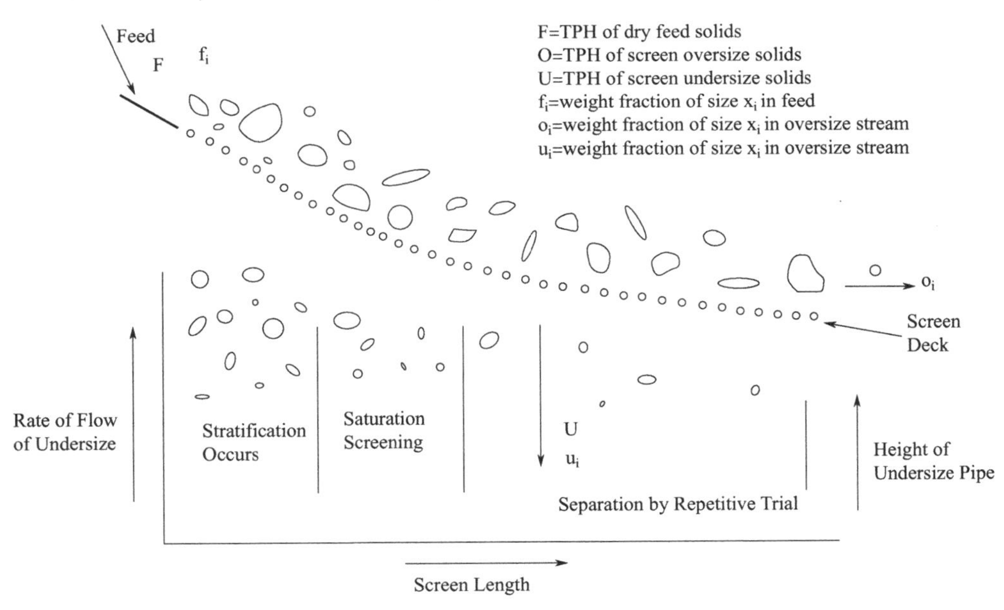

Figure 5.4 Variation in the mass falling through a screen along its length

Example 3:

Given: Use the screen separation information in Example 1, calculate the efficiency of undersize recovery.

From Equation (5-3)

$$E_r = 81/90 \times 100 = 90\%$$

or from Equation (5-4), a = 90%,

$$b = (90 - 81)/19 \times 100\% = 9/19 \times 100\% = 47\%$$

$$E_r = (90 - 47)100/90(100 - 40) \times 100\% = 90\%$$

Example 4:

Given: Use the information provided in example 2, calculate the efficiency of undersize recovery.

From Equation (5-3)

$$E_r = 45/50 \times 100\% = 90\%$$

or from Equation (5-4), a = 50%

$$b = (50 - 45)/55 \times 100 = 5/55 \times 100 = 9\%$$

$$E_r = (50 - 9)100/50(100 - 9) \times 100 = 90\%$$

In some instances, the customers may consider both the oversize and undersize (through) from the same screen as products. In this case, check both " Efficiency of Undersize Removal" and "Efficiency of Undersize Recovery." If there is a small amount of oversize, an efficiency of undersize removal looks poor, the screen area should be increased by 20%.

5.2 Mass (Material) Balances

Material quantities, as they pass through processing operations, can be described by material balances. Such balances are based on the conversion of mass.

Material balances are very important in material processing. It is fundamental to the control of processing, particularly in the control of yields of products. The first material balance is determined in the exploratory stage of a new process, improved during pilot plant experiments when the process is being planned and tested, checked out when the plant is commissioned, and then refined and maintained as a control instrument as production continues. When any changes occur in the process, the material balances need to be determined again.

In a unit operation, when its nature is seen as a whole or may be represented diagrammatically as a box (Figure 5.5), the mass going into the box must balance with the mass coming out.

Solid and liquid mass balances:

Node 1: Overall Circuit

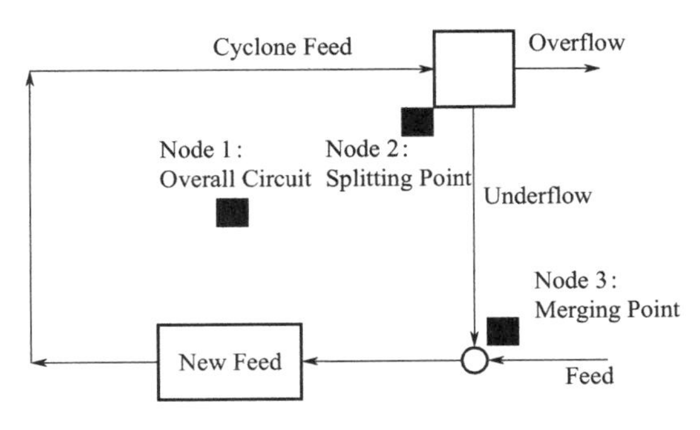

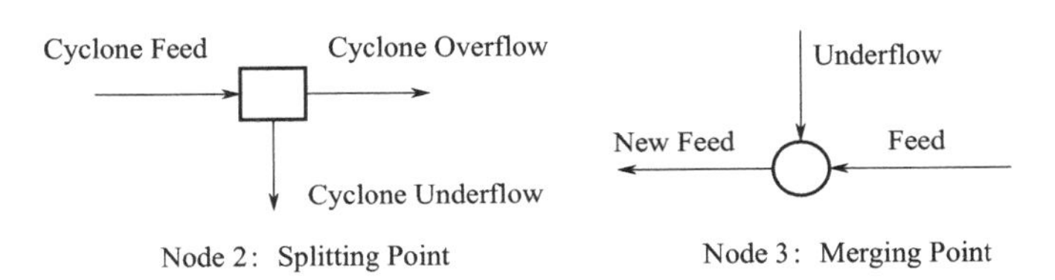

Figure 5.5 Mass balances in a closed grinding classified circuit

Overflow = Feed

O=F

Node 2: Splitting Point

Cyclone feed = Overflow + Underflow

CF = O + U

Node 3: Merging Point

Feed to Ball Mill:

New feed = Feed + Underflow

NF = F + U

Example: Material balances for a classified cyclone and ball mill circuit are shown in Figure 5.6.

The basic formulas of liquid and solid mass balances are defined as:

$$\text{USgpm} = \frac{\text{TPH} \times 4.0}{\text{SG}} \tag{5-5}$$

$$\text{solid\%} = \frac{\text{solid}}{(\text{solid} + \text{liquid})} \times 100\% \tag{5-6}$$

Table 5.3 lists the material balances for cyclone overflow, underflow, and feed. Table 5.4 lists the material balances for the ball mill.

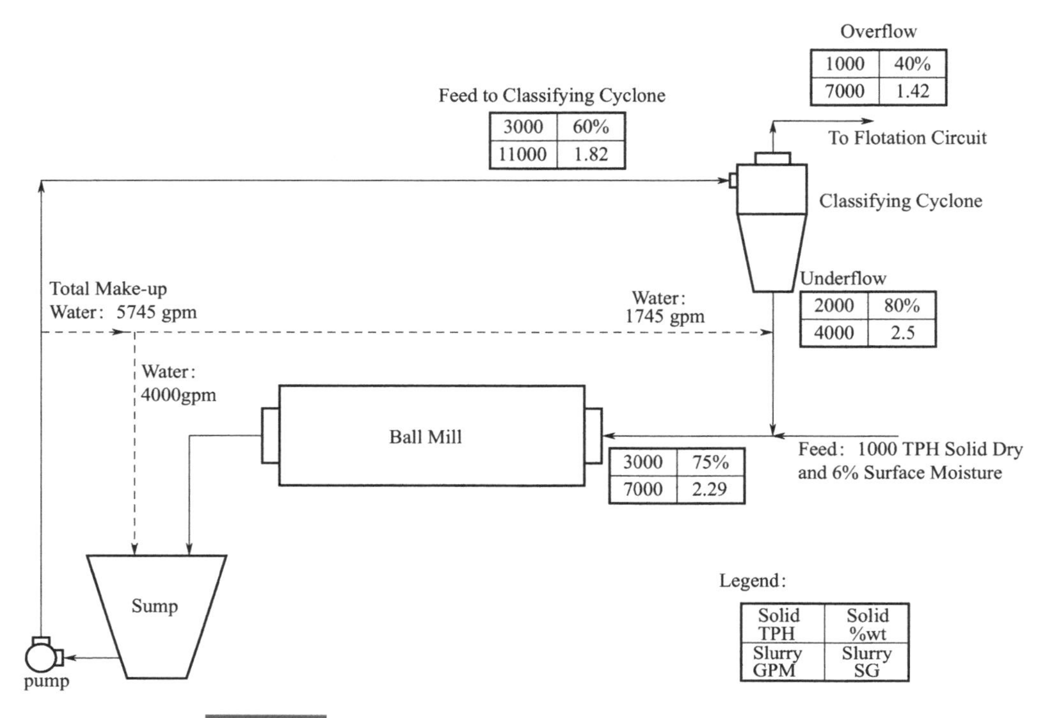

Figure 5.6 A closed grinding and classified cyclone circuit

Table 5.3 Material balance for the classified cyclone

Cyclone Overflow	wt, TPH	SG	Volume, USgpm
Solid	1000	4	1000
Liquid	1500	1	6000
Overflow Slurry	2500	1.42	7000
Percentage Solid	40 by wt.		14.29 by vol.
Cyclone Underflow:			
Solid (200% RCL)	2000	4	2000
Liquid	500	1	2000
Underflow Slurry	2500	2.5	4000
Percentage Solid	80① by wt.		50 by vol.
Cyclone Feed:			
Solid	3000	4	3000
Liquid	2000	1	8000
Cyclone Feed Slurry	5000	1.81	11000
Percentage Solid	60 by wt.		27.27 by vol.

① Water mass flow in the underflow slurry is not given. It is determined by calculating the liquid requirement to meet 60% solid by weight for the cyclone feed, 2000 TPH, first, and then, subtracted by 1500 TPH water in cyclone overflow to obtain 500 TPH liquid in underflow.

Table 5.4 Material balance for feed to ball mill

	wt, TPH	SG	Volume, USgpm
Solid	3000	4.00	3000
Liquid	1000	1.00	4000
Ball Mill Feed Slurry	4000	2.29	7000
Solid %	75 by wt.		14.29 by vol.

In Figure 5.3, the make-up water to the circuit is calculated as:

① When water needed in ball mill = 4000 gpm (or 1000 TPH) and water from the underflow of cyclone = 500 TPH

Calculation of the make-up water underflow of the cyclone and to new feed is as follows:

Assumed the dry feed has 6% surface moisture:

$$0.06=\frac{\text{TPH water}}{\text{TPH water}+1000\ \text{TPH solid}} \tag{5-7}$$

Therefore, the amount of water in the feed = $1000\ \text{TPH solid}\times\left(\frac{0.06}{1-0.06}\right)=63.83\text{TPH}$

Make-up water needed = total water needed − water from hydro cyclone underflow − water from surface moisture

$$=1000\text{TPH}-500\text{TPH}-63.83\text{TPH}=436.17\text{TPH}=\frac{436.17\text{TPH}\times 4}{1.00}=1744.68\text{gpm}\approx 1745\text{gpm}$$

② When water needed for cyclone feed = 8000 gpm and water from the ball mill = 4000 gpm,

the make-up water to sump (Feed Box) is:

$$8000\ \text{gpm} - 4000\ \text{gpm} = 4000\ \text{gpm}$$

Calculations of the slurry or pulp density can follow the formulae and example below.

③ Pulp density calculations:

Density of water (SG = 1.00):

$$\frac{1\text{g}}{\text{cm}^3}=\frac{1000\text{kg}}{\text{m}^3}=\frac{1000\text{kg}}{\text{m}^3}=\frac{1000\text{kg}\times 2.205\text{lb}/\text{kg}}{(3.281\text{ft})^3}=62.43\frac{\text{lb}}{\text{ft}^3}$$

$$=62.43\frac{\text{lb}}{\left(1\text{ft}^3\right)\left(\frac{7.481\text{gal}}{\text{ft}^3}\right)}=8.345\text{lb}/\text{gal}$$

Examples:

① 12 gpm of water (SG =1.00)

$$12\text{gpm}\times\left[8.345\frac{\text{lb}}{\text{gal}}\times(1.00)\times\frac{1}{\frac{\text{hr}}{60}\text{min}}\times\frac{1\text{ton}}{2000\text{lb}}\right]=12\times(1.00)\times\left[\frac{1}{4}\text{TPH}\right]=3\text{TPH}$$

② 4 gpm solid (SG = 2.00)

$$4\text{gpm}\times(2.00)\times\left[8.345\frac{\text{lb}}{\text{gal}}\times\frac{1}{\frac{\text{ht}}{60}\text{min}}\times\frac{1\text{ton}}{2000\text{lb}}\right]=4\times(2.00)\times\left[\frac{1}{4}\text{TPH}\right]=2\text{TPH}$$

③ Slurry or pulp (SG = X):

Since slurry in volume: 12 gpm + 4 gpm = 16 gpm, and slurry in weight: 3 TPH + 2 TPH = 5 TPH,

the SG of slurry is calculated by:

$$16\times(\text{X})\times\left[\frac{1}{4}\text{TPH}\right]=5\text{TPH}$$

solve for X: $\text{X}=\frac{5\times4}{16}=1.25$

the density of slurry is then: $8.345\frac{\text{lb}}{\text{gal}}\times1.25=10.425\frac{\text{lb}}{\text{gal}}$

Therefore, the specific gravity of the slurry is 1.25, and density is 10.425 lb/gal.

5.3 Crushing and Grinding Circuits

The comminution of as-mined coal generally involves three stages: ① primary breaking, ② secondary crushing, and ③ screening crushing. Designing selecting equipment for a three-stage layout is to provide optimum efficiency and high yields of marketable coal for a majority of coal preparation plants. The reduction ratio is one criterion applied to the choice of breakers and crushers and is usually within the range of 4 to 10.

① Primary crushing (Figure 5.7)

② Secondary crushing (Figure 5.8)

Basically, there are two types of crushing/grinding circuits: (I) open circuit and (II) closed circuit. They are described below:

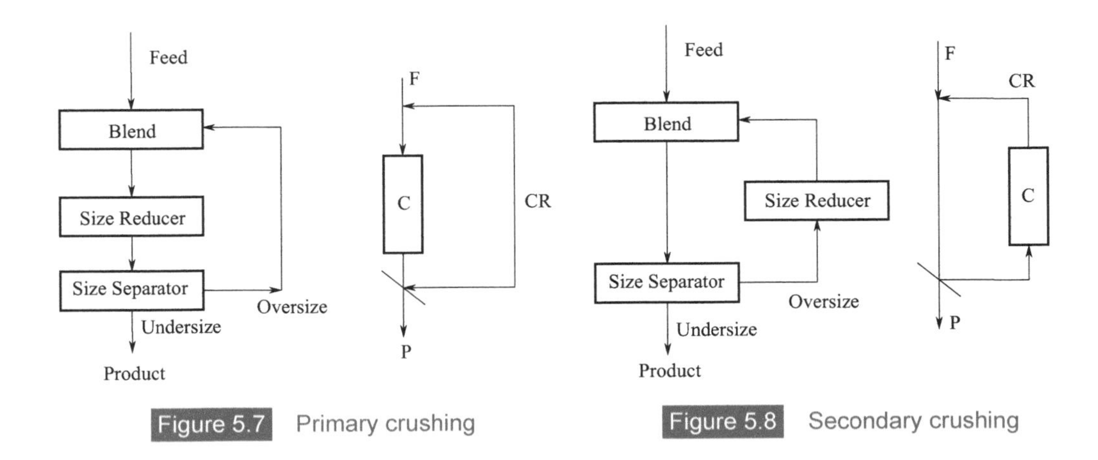

Figure 5.7 Primary crushing

Figure 5.8 Secondary crushing

5.3.1 Open Circuit

Open crushing circuit does not recirculate oversize product from screen to crusher. Examples are shown in Figures 5.9 and 5.10 for Bradford Rotary Breaker and the combination of a double roll crusher and a screen circuit, respectively.

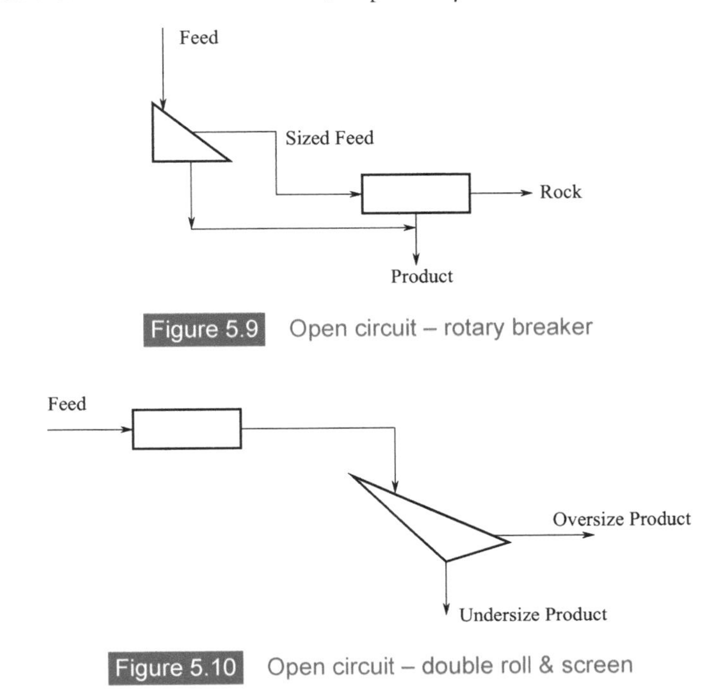

Figure 5.9 Open circuit – rotary breaker

Figure 5.10 Open circuit – double roll & screen

In the open circuit, material balance for the crusher at a steady state is:

$$F=P \tag{5-8}$$

Where,

F = Crusher feed

P = Product from crusher and undersize product from screen

5.3.2 Closed Circuit

A crushing machine does not reduce all their feed to a size equal to or smaller than the crusher setting. In normal practice, the crusher is set at a smaller setting (opening) than the screen desk opening, to obtain a good balance between the net finished product (100% minus product top size) and circulating load. By definition, the circulating load is the total feed to the crusher minus the original crusher feed.

With each pass through the crusher, an additional amount of oversize will be reduced to undersize. The number of passes for a given batch of material, with no addition of new feed, before 100% reduction is achieved, is infinite. The percentage of oversize remaining after each pass through the crusher and over the screen deck diminished in geometric progress.

(1) Regular closed circuit

A closed crushing circuit is shown in Figure 5.11. It has 30% oversize in the crusher product and 90% screen efficiency. Assuming the original feed to the crusher is sized, and it contains no fines or materials of lesser size than the desired crusher product, calculate the circulating load to the crusher. The geometric progress of crusher oversize is:

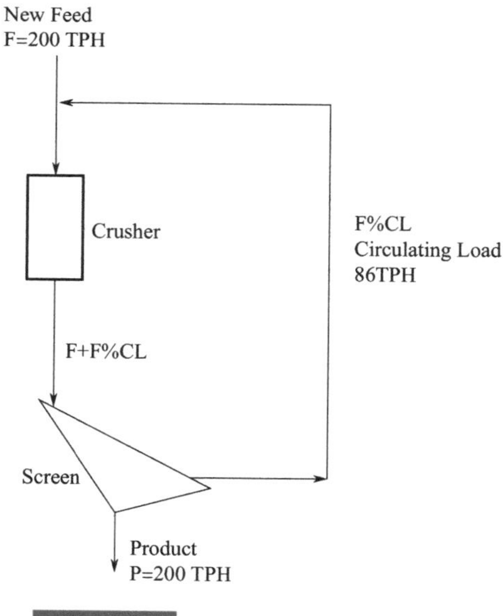

Figure 5.11 A regular closed circuit

Pass through crusher | TPH

$$200\times\left(0.3\times\frac{1}{0.9}\right)=66.67$$

$$200\times\left(0.3\times\frac{1}{0.9}\right)^2=22.22$$

$$200\times\left(0.3\times\frac{1}{0.9}\right)^3=7.41$$

$$200\times\left(0.3\times\frac{1}{0.9}\right)^4=2.47$$

$$200\times\left(0.3\times\frac{1}{0.9}\right)^5=0.82$$

99.86 TPH

−100 TPH

The circulating load can be written as

$$F\left(\frac{r}{E_m}\right)+F\left(\frac{r}{E_m}\right)^2+F\left(\frac{r}{E_m}\right)^3+F\left(\frac{r}{E_m}\right)^4+\cdots+F\left(\frac{r}{E_m}\right)^{n+1}=F\left(\frac{r}{E_m}\right)\left[\frac{\left(\frac{r}{E_m}\right)^n-1}{1-\left(\frac{r}{E_m}\right)}\right] \tag{5-9}$$

where

F = Original feed to the screen (TPH)

E_m = % Screen efficiency of undersize removal

R = % Oversize in crusher product

When $n\to\infty$, then $\left(\frac{E_m}{r}\right)^n\to 0$

Hence the circulation load is $F\left[\frac{1}{\left(\frac{E_m}{r}-1\right)}\right]$ (5-10)

From the equation above, the % circulating load is:

$$CL(\%)=\frac{1}{\left(\frac{E_m}{r}\right)-1}\times 100\%=\frac{1}{\left(\frac{90}{30}\right)-1}\times 100\%=50\% \tag{5-11}$$

Circulating load = 200 TPH × 50% = 100 TPH

In a closed grinding circuit, it gives 50% oversize in the grinding mill product and 75% classifying cyclone efficiency. Assume the original feed to the grinding mill is sized, calculate the % circulating load and circulating load to the mill. The % circulating load is:

$$CL(\%) = \frac{1}{\left(\frac{E_m}{r}\right) - 1} \times 100\% = \frac{1}{\left(\frac{75}{50}\right) - 1} \times 100\% = 200\%$$

Circulating load = 200 TPH × 200% = 400 TPH

(2) Reversed closed circuit

In practice, the original feed to the crusher is frequently not sized. As shown in the closed crusher circuit in Figure 5.12, it means that the feed would have to be screened before feeding the crusher. There is 20% oversized in the new feed. It gives 30% oversize in crusher product and 90% screen efficiency.

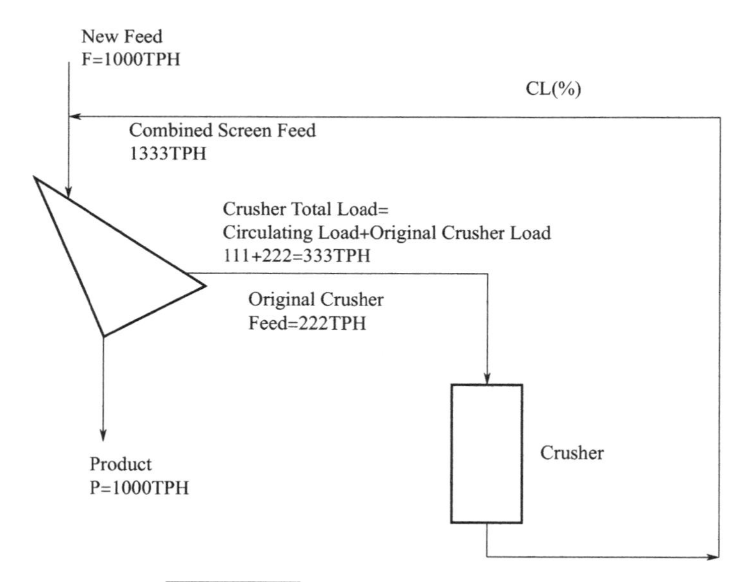

Figure 5.12 A reversed closed circuit

$$\text{Original Crusher Feed} = \text{Feed} \times \frac{\text{Oversize in original Feed}}{\text{Efficiency of Undersize Removal}} = F \times \frac{r_1}{E_m} \tag{5-12}$$

$$= 1000 \times \frac{0.2}{0.9} = 222\text{TPH}$$

$$\text{Circulating Load} = \text{Original Crusher Feed} + \text{Original Crusher Feed} = \text{F} \times \frac{r_1}{E_m} \%\text{CL}$$

$$= \text{F} \cdot \frac{r_1}{E_m} \frac{1}{\left(\frac{E_m}{r_2} - 1\right)} = 222 \times \frac{1}{(90/30) - 1} = 222 \times 0.5 = 111\text{TPH} \tag{5-13}$$

$$\text{Combined Feed to Screen} = \text{New Feed} + \text{Crusher Total Load} = 222 + 111 = 333\text{TPH}$$

where

F = Original feed to the screen (TPH)

E_m = % screen efficiency of undersize removal

r_1 = % oversize in feed

r_2 = % oversize in crusher product

Note:

For a grinding circuit, a crusher will be replaced by a grinding unit and a screen will be replaced by a classified cyclone.

5.3.3 Relationship between E_m and E_r

According to the definition adopted by the Vibrating Screen Manufacture Association (VSMA) for screen efficiency, screen efficiency of undersize recovery E_r is used as defined in Equation (5-3) or Equation (5-4). E_r is the screen efficiency of undersize recovery. In case E_r is available instead of E_m for the closed crushing circuit analysis, E_r can be converted into E_m by using the following equations:

$$E_m = \frac{\text{Theoretical Oversize}}{\text{Actual Oversize}} = \frac{F_r}{F - u} \tag{5-14}$$

$$E_c = \frac{\text{Actual Undersize}}{\text{Theoretical Undersize}} = \frac{100u}{F\left(1 - \frac{r}{100}\right)} \tag{5-15}$$

where

E_c is % screen efficiency of undersize recovery;

E_m is % screen efficiency of undersize removal;

F is original feed to the screen (TPH);

r is % oversize in crusher products;

u is actual undersize (TPH).

$$u = \text{Actual Undersize}(\text{TPH}) = \text{F} \times \frac{E_c}{100}\left(1 - \frac{r}{100}\right) \tag{5-16}$$

Substituting the E_m and E_c to the equation for Circulating Load, the following equation can be derived:

$$E_m = \frac{F_r}{F[1 - \frac{E_c}{100}\left(1 - \frac{r}{100}\right)} = \frac{r}{\left[1 - \frac{E_c}{100}\left(1 - \frac{r}{100}\right)\right]} \tag{5-17}$$

5.4 Types of Crushers Used in Mineral Processing

General crushing and milling equipment used in mineral processing are listed as follows.

(1) Rotary breaker

The Rotary Breakers crush materials by gravity impact only. A large cylinder made of perforated screen plates is fitted with internal shelves. As the cylinder rotates, the shelves lift the feed and, in turn, the feed slides off the shelves and drops onto the screen plates below, where it shatters along natural cleave lines. The over-sized pieces will again be lifted and dropped by the shelves until they too pass through the screen plates. The size of the screen plate perforations determines the product size. The uncrushable debris and rocks flow to the discharge end of the cylinder.

(2) Single and double roll crusher

It reduces a large feed by a combination of shear, impact, and compression. The Hercules single-roll crushers are noted for low headroom requirements and large capacities.

(3) Hammer mill

When a non-reversible hammermill is used for reduction, material is broken first by impact between hammers and material and then by a scrubbing action (attrition) of material against screen bars.

(4) Impact crusher

The bottom of the reversible impactor is open and the sized material passes through almost instantaneously. A liberal clearance between hammers and the breaker blocks eliminates attrition, and crushing is by impact only.

(5) Jaw crusher

The jaw crushers crush by compression without rubbing. Hinged overhead and on the centerline of the crushing zones, the swinging jaw meets the material firmly and squarely. There is no rubbing action to reduce capacity, generate fines, or cause excessive wear of jaw plates.

5.5 Unit Operations for Screening, Classification, and Desliming

Unit operations for screening, classification, and desliming include grizzly scalper, screens (vibration and stationary types and high-frequency banana types), sieve bend, hydro cyclone classifier and spiral classifiers (horizontal rake or screw types).

(1) Grizzly scalper screens

Figure 5.13 is a grizzly deck screen. During operation, the grizzly screen vibrates slowly, possibly in combination with spraying water. The feed material passes through several stages of the screen. Oversize material is screened out and then returned for crushing.

Figure 5.13 Grizzly deck screen

The characteristics of a standard grizzly scalper include:

① High-volume fines removal.

② Moves material faster and smoother.

③ Compact design fits in tight quarters.

④ Adjustable cast manganese bars.

⑤ Available in 3′×4′ and 4′×4′ sizes.

(2) Vibrating screen

Vibrating screens are the screens of choice in modern practice. The range of screen aperture sizes is from 10 inches to 100 mesh (254 to 0.15 mm). Screen slope will vary from 16° to 26° . The stroke depends upon the size of separation and varies from ½ to 1/32 inch (12.7 to 0.8 mm). Vibration frequency ranges from 660 Hz to 3400 Hz for the finest separations.

Figure 5.14 shows a vibrating screen used in industrial practice.

The vibration screen operates at a high frequency to move the feed material travelling on the screen deck. The high frequency of vibration screens also increases the chances of contact between particles and screen openings. Undersize material passes through the screen holes, while the oversize material exits from the end of the screen deck.

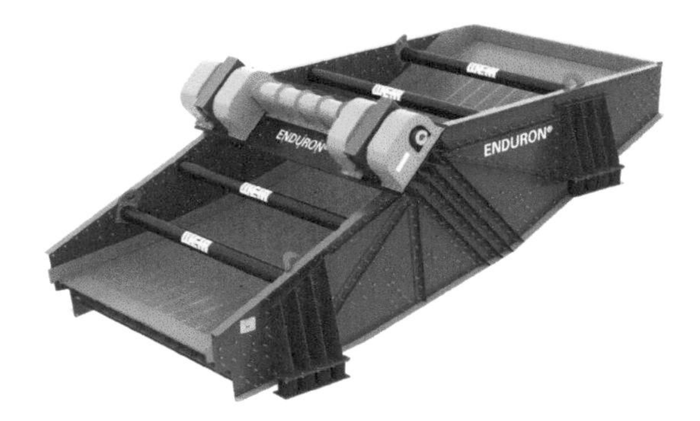

Figure 5.14 Weir group's Enduron low-head banana screen

(3) Sieve bend

Sieve bends are used both for sizing and dewatering service. The sieve bend aperture is from 0.079 to 0.0049 inch (2 to 0.125 mm) and the width is from 1 to 5 ft (0.305 to 1.52 m). It has no moving parts with low operating costs and has increasing use in fine coal circuits. Figure 5.15 illustrates the sieve bend.

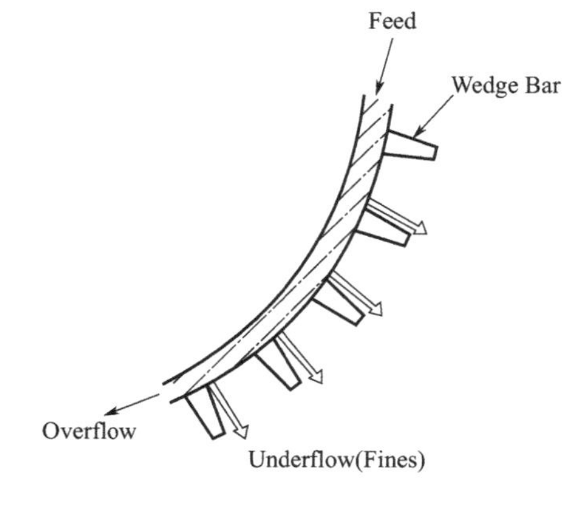

Figure 5.15 Sieve bend

(4) Screen decking on static sieves

Bee-Zee screens are used in static sieves in all industries in either classifying or dewatering slurry flows by gravity over an inclined screen surface. Screen rods perpendicular to flow slice away a layer of slurry. Separation is much smaller than actual screen openings – usually about one-half. Other important points – the sieve bend, as shown in Figure 5.15, needs turbulent-free flow. The feed material needs proper percent of solids. In case of wear of the screen deck, rotate the screen surface 180 degrees to position the sharp high leading edge of profiles towards slurry flow. The dull trailing edge is resharpened during operation. Figure 5.16 shows the schematic drawing of a screen surface.

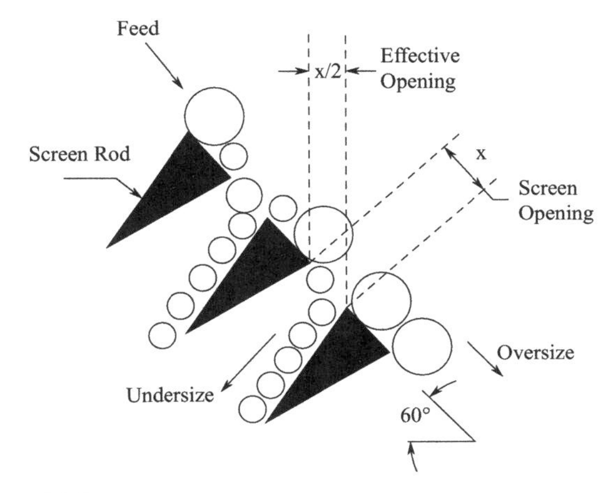

Figure 5.16 Screen surface of screen decking on static sieves

5.6 Classification by Hydrocyclone

The liquid-solid cyclone (hydrocyclone) was introduced by Driessen in 1939 (Tangel and Brison, 1956). Its application includes (1) classification or sizing of particles, degritting liquid, or liquid suspensions of fine solids; (2) when desirable, the separation of objectionable granular materials; and (3) desliming operations.

The cyclone operates by centrifugal force generated by feeding the cyclone tangentially, forcing the liquid and solids to the wall of the cyclone. The centrifugal force increases the settling rate of the particles so that the faster settling coarser particles are forced to the wall and are discharged to the underflow of the cyclone. The finer solids do not have sufficient time to migrate to the underflow and are discharged to the overflow through the vortex finder. Figure 5.17 is a cyclone and cyclone nomenclature. Figure 5.18 is a hydrocyclone cluster. A hydrocyclone cluster is a piece of equipment that allows for multiple hydrocyclones to operate alongside one another in a single unit. They are typically engineered to order and customized to fit into the space available.

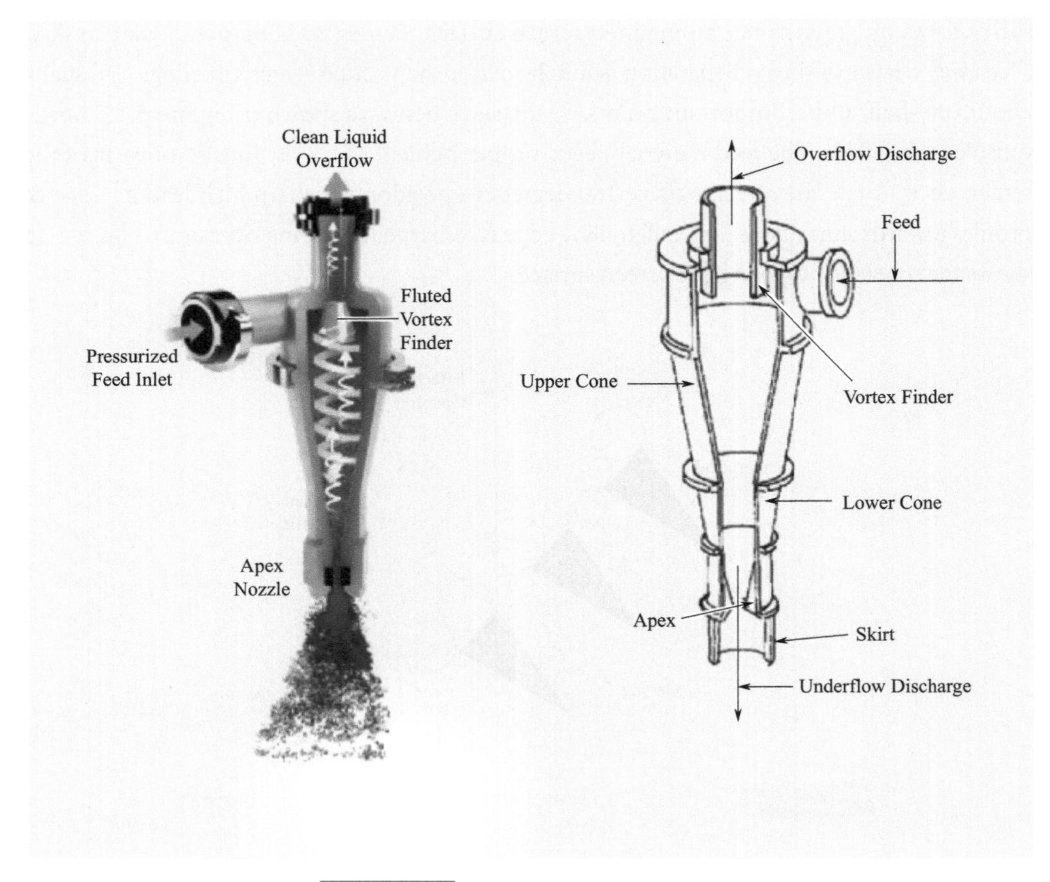

Figure 5.17 Classifying hydrocyclone

Figure 5.18 Cluster of hydrocyclones

WASHABILITY ANALYSIS

The float-and-sink (Washability) tests are conducted on samples of coal from coalbeds that will likely need washing (beneficiation, preparation) to remove sulfur and/or ash (rock and mineral matter) from the coal to meet the desired specifications of the end users. The washability analysis determines the amount of coal that can be separated from rock and minerals in heavy medium suspensions of different densities (Kentucky Geological Survey).

6.1 Float-and-Sink Test

6.1.1 Heavy Medium Suspensions for Float-and-Sink Tests

The solution used in the float-and-sink tests, known as heavy medium suspensions, is lithium tungstate (LMT), in which the tungsten anion is $[H_2W_{12}O_{40}]^{6-}$.

The specific gravity of the LMT solution is approximately 2.85~3.7, and the pH is 6. The high specific gravity supersaturated solution has no odor and appears pale yellow in color.

The LMT solution is not discarded but recycled and reused. The contaminant in used solutions is coal dust. The coal dust particles in the solution can be removed by vacuum filtering. The LMT solution should be stored in the same container, or a new glass container, after filtering and readjusting back to the specified specific gravity.

Examples: To prepare a medium of specific gravity (SG) of 1.8, how much 2.9 SG medium and deionized water are needed?

Solution:

Assuming the volume of the required deionized water is X and using a total volume of 1 ml as a basis. Applying the mass balance rule:

1 ml (1.8 g/ml) = X ml (1.0 g/ml) + (1 − X) ml (2.90 g/ml)

1.80 = 1.0 X + 2.90 − 2.90 X

(2.90 − 1.0) X = 2.90 − 1.80

Therefore

X = 0.579 ml for 1.0 SG deionized water

1 − X = 1 − 0.579 = 0.421 ml for 2.90 SG LMT

The ratio of medium to water is 0.421 ml / 0.579 ml = 0.727: 1 (vol/vol)

6.1.2 Float-and-Sink Test Procedures

There are specially designed apparatus for use in a float-and-sink test. They are various sizes and shapes, as shown in Figure 6.1. For the smaller-scale laboratory units, they are regular flasks and funnels with ground joints made of Pyrex glass. A selected heavy medium suspension with a specific gravity is filled up half-way of the upper funnel. Coal samples are added to the solution and stirred with a glass rod until well blended. A rubber stopper with a stainless-steel handle is used to separate the float and sink products. The upper funnel is then separated from the flask. The coal and the solutions in the flask and funnel are poured into the Buchner funnels separately for vacuum filtration.

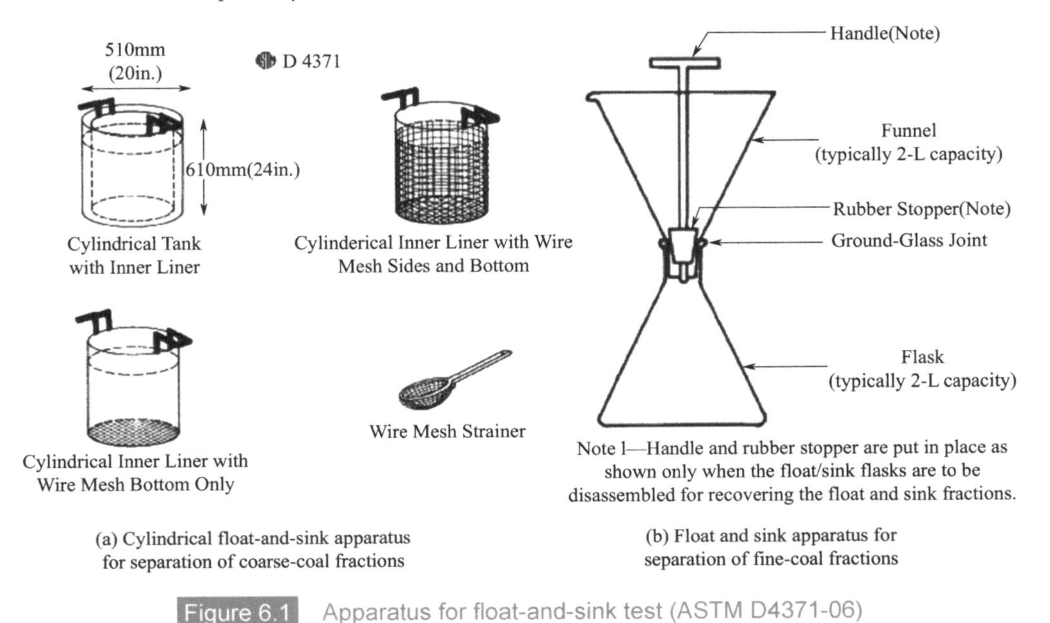

Figure 6.1 Apparatus for float-and-sink test (ASTM D4371-06)

6.2 Washability Data Analysis

Washability data obtained from the float-and-sink tests for a given raw coal is shown in Table 6.1 which includes the specific gravity analysis for (individual fraction) wt%, ash%, sulfur%, and heating value (Btu/lb). To establish the washability curves for the raw coal, the cumulative float and cumulative sink data must be calculated. The washability curves will provide important information about the theoretical yields and quality of clean coal and refuse products, as well as the degree of difficulty in washing the raw coal. The cumulative float data are obtained by summing up from the lightest fraction to the heaviest fraction, while the cumulative sink data are obtained by summing up from the heaviest fraction to the lightest fraction. Other important correlations of washability data are also shown below.

If the raw coal is to be cleaned at 1.60 specific gravity of separation (SG50) or cut point, then the theoretical values can be calculated as follows:

① Cumulative float (%) = 46.5 + 38.1 + 7.2 = 91.8(%)

Table 6.1 Washability analysis of raw coal

Specific Gravity Interval	Direct				Cumulative Float					Cumulative Sink				Ordinate Z
	wt %	Ash %	Sul %	Btu/lb	wt %	Ash %	Sul %	Btu/lb	SO_2/mmBtu	wt %	Ash %	Sul %	Btu/lb	
< 1.30	46.5	6.6	1.0	14051	46.5	6.6	1.00	14051	1.42	100.0	15.4	2.36	11900	23.2
1.30~1.40	38.1	12.4	2.2	12514	84.6	9.2	1.54	13359	2.31	53.3	23.1	3.54	10032	65.5
1.40~1.60	7.2	27.1	5.3	5746	91.8	10.6	1.84	12762	2.88	15.4	49.7	6.87	3890	88.2
1.60~1.80	1.8	47.4	7.8	2811	93.6	11.3	1.95	12570	3.10	8.2	69.5	8.19	2260	92.7
> 1.80	6.4	75.7	8.3	2105	100.0	15.4	2.36	11900	3.97	6.4	75.7	8.30	2105	96.8
Total	100.0													

② Cumulative float,

ash% = (6.6 × 46.5 + 12.4 × 38.1 + 27.1 × 7.2)/ (46.5 + 38.1 + 7.2) × 100% = 10.6%

③ Cumulative float,

Sulfur% = (1.0 × 46.5 + 2.2 × 38.1 + 5.3 × 7.2)/ (46.5 + 38.1 + 7.2) × 100% = 1.84%

④ Cumulative float,

Heating value, Btu/lb = (14051 × 46.5 + 12514 × 38.1 + 5746 × 7.2)/ (46.5 + 38.1 + 7.2) = 12,762 Btu/lb

⑤ $S + O_2 = SO_2$ (molar mass of sulfur is 32 g/mol, and oxygen is 16 g/mol)

S/mmBtu = (1 lb coal × 1.84%)/(1 lb coal × 12762 Btu/lb × mmBtu/10^6 Btu)

=1.44 lb S/mmBtu

SO_2/mmBtu = (1 lb coal × 1.84% × (32+16×2)/32) / (1 lb coal × 12762 Btu/lb × mmBtu/10^6 Btu)

=2.88 lb SO_2/mmBtu

⑥ Cumulative sink (%) = 6.4 + 1.8 = 8.2%

⑦ Cumulative sink, ash % = (75.7 × 6.4 + 47.4 × 1.8)/ 8.2 = 69.49%

⑧ Cumulative sink, sulfur % = (8.3 × 6.4 + 7.8 × 1.8)/ 8.2 = 8.19%

⑨ Cumulative sink, Btu/lb = (2105 × 6.4 + 2811 × 1.8)/ 8.2 = 2260 Btu/lb

⑩ Elementary ash curve: The elementary ash curve is designed to show the highest ash content of any individual particle that may be found in the float-coal product at any specific gravity. As illustrated in Figure 6.2, the elementary ash curve is established by finding a new ordinate Z and plotting the new ordinate values against the individual ash value of each specific gravity fraction as the abscissa. The value of ordinate Z is found by solving the equation

$$Z_i = \sum_{i=1}^{n} w_{i-1} + \frac{w_i}{2} \tag{6-1}$$

where $\sum_{i=1}^{n} w_{i-1}$ is the cumulative weight percent float of all materials of lower than the ith specific gravity; w_i is the weight percent of the material at the given specific gravity. Steep slopes represent relatively small ash differences for large differences in yield, whereas a flat slope indicates easy separation.

The highest ash content of the particle at specific gravity of 1.40-1.60 is 27.1%. To calculate the highest ash content of the particle at any specific gravity interval, follow the example given below. At specific gravity of 1.45, the cumulative float, wt% is 88% and the average ash content is 9.5% (linearly interpolated from Table 6.1). At the specific gravity of 1.40, the cumulative float, wt% is 84.6% and the average ash content is 9.2%. The individual fraction of ash content at 1.40-1.45 can be calculated as follows:

Ash% at specific gravity 1.40 − 1.45 = (9.5 × 88 − 9.2×84.6) / (88-84.6) = 17.0%

Z = 84.6 + [(88 − 84.6)/2] = 86.3%

Therefore, the highest ash content at specific gravity 1.4 is 17%.

Additional calculations to illustrate the use of Equation (6-1):

For SG <1.30 (n = 1)

(The values in the parenthesis are cumulative value which is represented by the summation symbol in the formula)

$Z_1 = w_0 + w_1 / 2 = (0) + 46.5 / 2 = 23.2$

For SG <1.40 (n = 2)

$Z_2 = (0 + w_1) + w_2 / 2 = (0 + 46.5) + 38.1 / 2 = 46.5 + 19.05 = 65.55$

For SG <1.60 (n = 3)

$Z_3 = (0 + w_1 + w_2) + w_3 / 2 = (0 + 46.5 + 38.1) + 7.2 / 2 = 84.6 + 3.6 = 88.2$

For SG <1. 80 (n = 4)

$Z_4 = (0 + w_1 + w_2 + w_3) + w_4 / 2 = (0 + 46.5 + 38.1 + 7.2) + 1.8 / 2 = 91.8 + 0.9 = 92.7$

For all materials (including SG < 1.80 and SG > 1.80 materials):

$Z_5 = (0 + w_1 + w_2 + w_3 + w_4) + w_5 / 2$

$= (0 + 46.5 + 38.1 + 7.2 + 1.8) + 6.4 / 2 = 93.6 + 3.2 = 96.8$

(This part is the cumulative float)

The results of ordinate Z are shown in Table 6.1.

⑪ Determination of the degree of difficulty in coal washing

The percentage of near gravity ± 0.10 material at the specific gravity of separation is used to determine the degree of difficulty in coal washing.

The amount of cumulative float, wt%, is obtained from washability data and cumulative float curve (see Figure 6.2). The difference between the two cumulative float yields, wt%, at the specific gravities of 0.1 higher or lower than a specific gravity of interest is computed first, and the difference is then divided by the cumulative float, wt%, at the specific gravity of 2.00. The + 0.1 specific gravity distribution curve is constructed by plotting the + 0.1 value, wt%, versus the specific gravity of interest. The data are shown in Table 6.2.

From the specific gravity curve, the cumulative float % is found as 92.5% and 90.0% at the specific gravity of 1.70 (=1.60 + 0.1) and 1.50(=1.60 − 0.1), respectively. Extending the specific gravity curve to the specific gravity of 2.00, the cumulative float % is found to be 94.0%. The ± 0.1 value for a specific gravity of 1.60 is exemplified below:

(92.5 − 90.0) / 94.0 × 100% = 2.7%

Comparing this ± 0.1 value of 2.7 percent with the values listed in Table 6.3, it shows that this raw coal is relatively simple to clean at 1.60 specific gravity.

Table 6.2 ± 0.1 specific gravity (SG) distribution curve

Specific gravity interval	Cumulative Float, %	± 0.10 Value, %	at SG
< 1.30	46.5		
1.30~1.35	76.0		
1.35~1.40	84.6	46.3	1.40
1.40~1.45	88.0	16.2	1.45
1.45~1.50	90.0	7.7	1.50
1.50~1.55	91.2		
1.55~1.60	91.8	2.7	1.60
1.60~1.70	92.5	1.9	1.70
1.70~1.80	93.6		
1.80~2.00	94.0		
< 2.00	100.0		

Table 6.3 Range of values of near-gravity materials

Quantity within 0.10 SG Range, %	Degree of difficulty in separation
0~7	Simple
7~10	Moderately Difficult
10~15	Difficult
15~20	Very Difficult
20~25	Extremely Difficult
above 25	Formidable

⑫ Washability curves

Washability curves are listed below and plotted in Figure 6.2. They include the following curves:

a. the cumulative float ash curve: plot cumulative float, wt% vs. cumulative float, ash%

b. the cumulative float sulfur curve: plot cumulative float, wt% vs. cumulative float, sulfur%

c. the cumulative float heating value curve: plot the cumulative float, wt% vs cumulative float heating value, Btu/lb

d. the cumulative sink ash curve: plot the cumulative sink, wt% vs. cumulative sink, ash%

e. the elementary ash curve (or characteristic curve): plot the ordinate Z vs. individual fraction, ash%

f. the specific gravity curve: plot the cumulative float, wt% vs. specific gravity

g. ±0.10 specific gravity distribution curve: plot 0.10 value wt% vs. specific gravity

Figure 6.3 illustrates the mass, Btu, ash, sulfur distributions vs. specific gravity and size of raw coal, and mass and Btu distributions vs. specific gravity and size of clean coal. After the raw coal has been cleaned, the materials of high specific gravity, high ash and sulfur contents, and low Btu have been discarded.

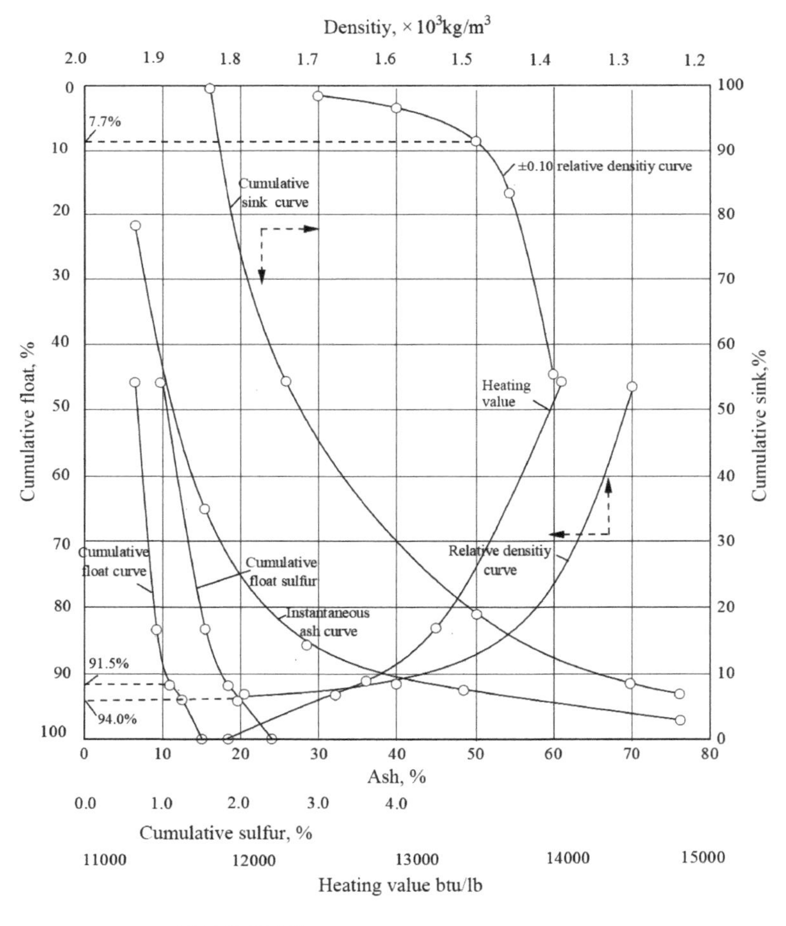

Figure 6.2 Raw coal washability curve

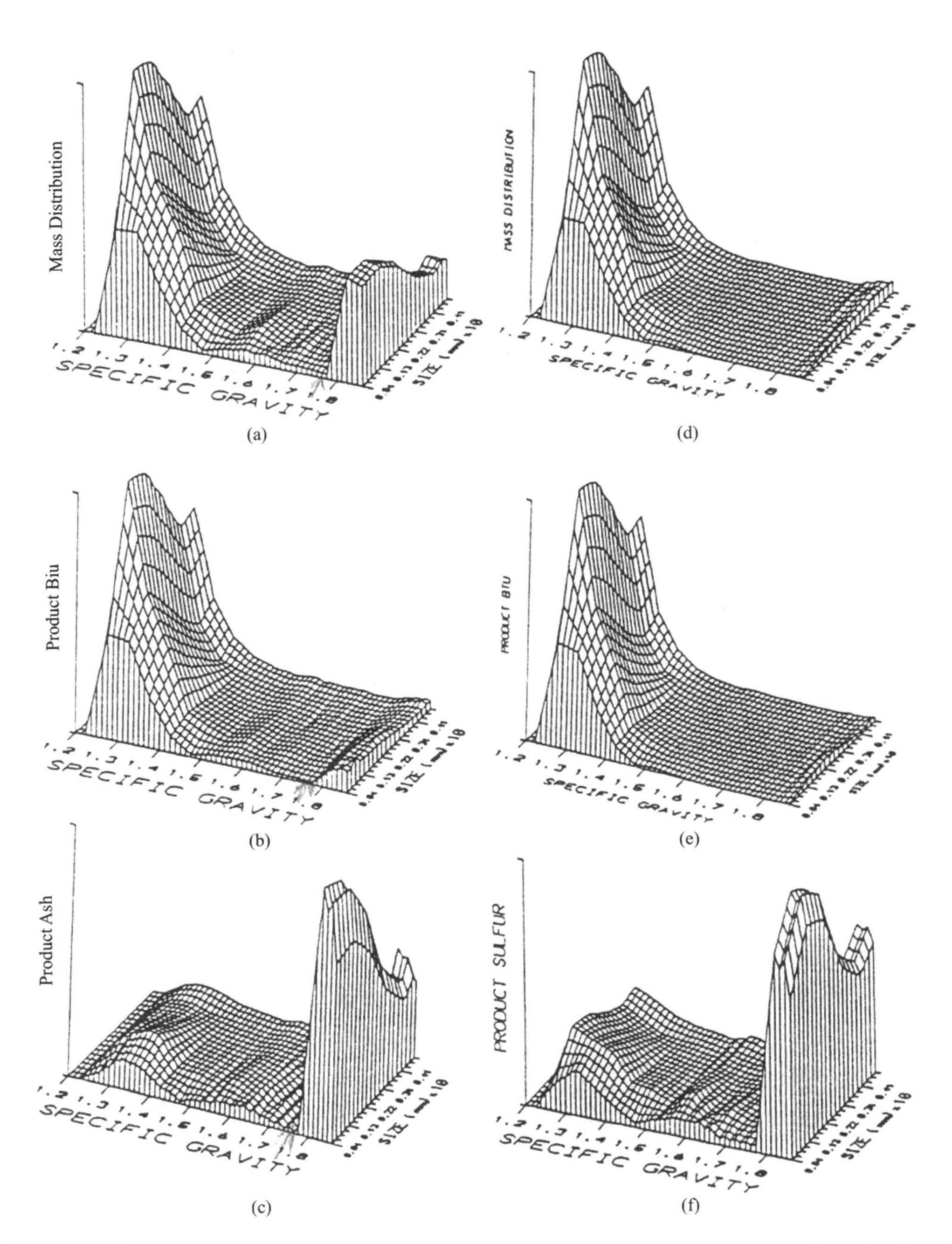

Figure 6.3 Response surfaces of raw coal washability data and clean coal products for Pond Creek seam coal

(a), (b), (c), (d) Raw coal; (e), (f) Clean coal product

6.3 Composite Washability Data for Coal

A set of raw coal washability data for coarse-size coal, intermediate-size coal, and fine-size coal is shown in Table 6.4~6.6, respectively. The composite washability data for the overall coal size range (-2-in), obtained from the weighted average of the percentage of feed of the washability data for coarse, intermediate, and fine-size coal, is shown in Table 6.7. The formula based on the weighted average method is used to calculate the composite sample, which is given in Table 6.8. The calculation for the composite washability for coal is tedious work. Thus, it is necessary to develop a computer program to accomplish this task.

Table 6.4 Washability data for coarse coal

(size range: -2-in + 1/4-in, weight:55.51% of feed (r_1))

Specific Gravity Interval	Individual components (dry base)			
	wt%(wj1)	Ash%(aj1)	Sulfur%(sj1)	Btu/lb(bj1)
< 1.30	39.34	4.37	0.83	14532
1.30~1.35	11.82	8.67	0.86	13728
1.35~1.40	12.55	12.86	0.84	13018
1.40~1.45	2.63	17.89	0.96	12217
1.45~1.50	2.99	22.22	1.33	11535
1.50~1.55	1.62	25.36	1.65	11018
1.55~1.60	1.14	29.33	1.95	10269
1.60~1.70	1.60	34.68	1.85	9379
> 1.70	26.31	77.53	1.77	2347

Table 6.5 Washability data for intermediate coal

(size: -1/4-in + No.30 US sieve, weight: 36.21% of feed (r_2))

Specific Gravity Interval	Individual components (dry base)			
	wt%(wj2)	Ash%(aj2)	Sulfur%(sj2)	Btu/lb(bj2)
< 1.30	53.90	3.52	0.79	14563
1.30~1.35	15.13	9.52	0.88	13619
1.35~1.40	3.63	13.25	0.97	13041
1.40~1.45	3.26	16.18	1.01	12545
1.45~1.50	2.61	19.09	1.19	11962

Specific Gravity Interval	Individual components (dry base)			
	wt%(wj2)	Ash%(aj2)	Sulfur%(sj2)	Btu/lb(bj2)
1.50~1.55	1.64	23.83	1.63	11183
1.55~1.60	1.29	27.87	1.90	10472
1.60~1.70	1.30	34.67	2.15	9372
> 1.70	17.24	78.25	2.00	2137

Table 6.6 Washability data for fine coal

(size: -No.30 US sieve, weight: 8.28% of feed (r_3))

Specific Gravity Interval	Individual components (dry base)			
	wt%(wj2)	Ash%(aj2)	Sulfur%(sj2)	Btu/lb(bj2)
< 1.30	24.27	1.65	0.74	14851
1.30~1.35	7.28	5.23	0.78	14216
1.35~1.40	4.13	9.38	0.8	13517
1.40~1.45	2.94	9.85	0.83	13433
1.45~1.50	4.15	12.28	0.84	13040
1.50~1.55	2.83	14.75	0.75	12643
1.55~1.60	6.96	19.06	0.65	11935
1.60~1.70	13.54	20.89	0.59	11582
> 1.70	33.91	52.93	1.79	6293

Table 6.7 Washability data for composite coal

(size: -2-in. weight: 100% of feed)

Specific Gravity Interval	Individual components (dry base)			
	wt%(wj2)	Ash%(aj2)	Sulfur%(sj2)	Btu/lb(bj2)
< 1.30	43.36	3.86	0.81	14561
1.30~1.35	12.64	8.87	0.86	13704
1.35~1.40	8.62	12.78	0.86	13041
1.40~1.45	2.88	16.51	0.97	12454
1.45~1.50	2.95	20.06	1.23	11847
1.50~1.55	1.73	23.39	1.52	11295
1.55~1.60	1.68	25.39	1.49	10898
1.60~1.70	2.48	28.44	1.34	10374
> 1.70	23.66	74.80	1.83	2760

Table 6.8 Example for calculating composite of samples (washability data)

Washability data for B seam coal

-2"+1" size fraction, 40% wt

Sp. gr. interval	Direct	
	wt%	ash%
<1.30	17	5
1.30~1.60	25	15
1.60~1.80	30	20
>1.80	28	30

rj1% = 40%

wt%	ash%
w11	a11
w21	a21
w31	a31
w41	a41
wj1	aj1

Washability data for B seam coal

-1"+1/4" size fraction, 60%wt

Sp. gr. interval	Direct	
	wt%	ash%
<1.30	26	7
1.30~1.60	33	18
1.60~1.80	27	26
>1.80	14	45

rj2% = 60%

wt%	ash%
w12	a11
w22	a21
w32	a31
w32	a41
wj2	aj2

Washability data for B seam coal

-2"+1/4" size fraction, 100%wt

Sp. gr. interval	Direct	
	wt%	ash%
<1.30	23.6	6.39
1.30~1.60	29.8	16.99
1.60~1.80	28.2	23.45
>1.80	19.6	36.43

r% = rj1 + rj2

wt%	ash%
w1	a1
w2	a2
w3	a3
w4	a4
wj	aj

Formula for calculating wt% and ash% for composite sample:

wj = (rj1*wj1 + rj2*wj2)/(rj1+rj2)

aj = (rj1*wj1*aj1+rj2*wj2*aj2)/(rj1*wj1+r2*wj2)

w1=	(40*17+60*26)/(40+60) =	23.6
w2 =	(40*25+60*33)/(40+60) =	29.8
w3 =	(40*30+60*27)/(40+60) =	28.2
w4 =	(40*28+60*14)/(40+60) =	19.6

a1 =	(40*17*5+60*26*7)/(40*17+60*26) =	6.39
a2 =	(40*25*15+60*33*18)/(40*25+60*33)	16.99
a3 =	(40*30*20+60*27*26)/(40*30+60*27)	23.45
a4 =	(40*28*30+60*14*45)/(40*28+60*14)	36.43

GRAVITY BASED SEPARATION PROCESSES

7.1 Fundamental of Momentum Transport and Settling Velocity of Particles in Fluid

Figure 7.1 illustrates the forces applied on a particle in a fluid. According to Newton's second law of motion,

$$F = ma \tag{7-1a}$$

$$F = F_w - (F_s + F_k) \tag{7-1b}$$

where

a = acceleration of motion of particle in fluid

F = force on a particle in fluid

F_k = drag force that acts in the opposite direction to that of motion

F_s = buoyant force exerted by fluid

F_w = weight of particle

m = mass of particle

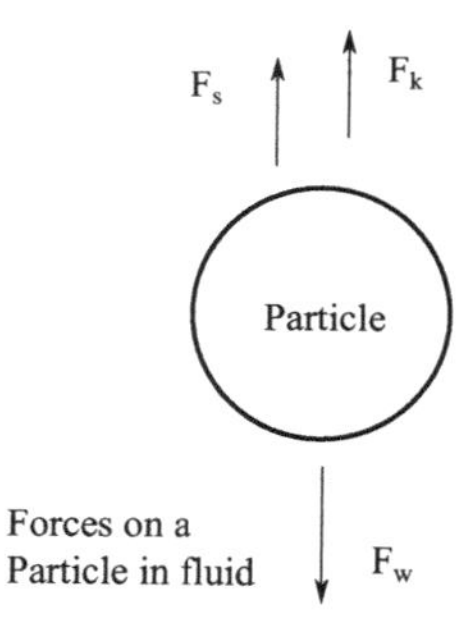

Figure 7.1 Forces on a particle in fluid

7.1.1 Criteria for Viscous Flow (Stokes' Criterion for Small Particles)

Small particles are referred to -1.68 mm (US sieve No.12). The settling velocity for those particles is derived below.

When the particle is small, or in a low Reynolds number region, the momentum is smaller than the friction force. Thus the drag force can be expressed as

$$F_k = 6\pi\left(\frac{D_p}{2}\right)\mu v \tag{7-2}$$

Reynolds number is $Re = \dfrac{D_p \delta \rho_s v}{\mu}$

Therefore,

$$F = ma = m\frac{dv}{dt} \tag{7-3a}$$

$$F = mg - \left(\frac{\rho_f}{\rho_s}\right)m - 6\pi\left(\frac{D_p}{2}\right)\mu v \tag{7-3b}$$

where

D_p = particle diametere　　ρ = specific gravity

μ = viscosity of fluid　　s = denote "solid"

δ = density of water, mg/l　　f = denote "fluid"

m = mass of particle　　v = velocity of particle

For fine particles, Equation (7-3a) = Equation (7-3b),

$$m\frac{dv}{dt} = mg - \left(\frac{\rho_f}{\rho_s}\right)m - 6\pi\left(\frac{D_p}{2}\right)\mu v$$

Divided by $m = \frac{4}{3}\pi\left(\frac{D_p}{2}\right)^3 \delta\rho_s$ on both sides, then

$$\frac{dv}{dt} = \frac{(\rho_s - \rho_f)\delta g}{\delta \rho_s} - \frac{18\mu v}{D_p^2 \delta \rho_s}$$

$$= \frac{(\rho_s - \rho_f)\delta g D_p^2}{18\mu} \times \frac{18\mu}{D_p^2 \delta \rho_s} - \frac{18\mu v}{D_p^2 \delta \rho_s}$$

$$= \frac{18\mu}{D_p^2 \delta \rho_s}\left[\left(\frac{D_p^2(\rho_s - \rho_f)\delta g}{18\mu}\right) - v\right]$$

$$= \frac{18\mu}{D_p^2 \delta \rho_s}(v_t - v)$$

$$= k(v_t - v) \tag{7-4}$$

Where　　$k = \frac{18\mu}{D_p^2 \delta \rho_s}$

$$\int_0^v \frac{dv}{v_t - v} = \int_0^t k\,dt$$

$$v = v_t(1 - e^{-kt}) \tag{7-5}$$

The terminal velocity is reached when the force of the system is in equilibrium, i.e., dv/dt = 0 and $v = v_t$. From Equation (7-4):

$$0 = \frac{(\rho_s - \rho_f)g}{\rho_s} - \frac{18\mu}{Dp^2\delta\rho_s} v_t$$

$$v_t = \left[\frac{D_p^2(\rho_s - \rho_f)\delta g}{18\mu}\right]^{1.0} \quad \text{for Stokes' Criteria} \tag{7-6}$$

$$v_t = \frac{D_p^2 \delta g}{18\mu}(\rho_s - 1) = k_s D_p^2(\rho_s - 1) \quad \text{for water } \rho = 1$$

where $k_s = \dfrac{\delta g}{18\mu}$

7.1.2 Criteria for Turbulent Flow (Newton's Criteria for Large Particles)

Larger particles are referred to those +1.68 mm (US sieve No.12). The settling velocity for those particles is derived below.

When a particle is large, the momentum is greater than the friction force. Thus, the drag force can be expressed as

$$F_k = C_D \pi \left(\frac{D_p}{2}\right)^2 \left(\frac{1}{2}\delta\rho_s v^2\right) \tag{7-7}$$

which is the drag force in the turbulent flow region.

Substituting Equation (7.7) into Equation (7-3b), it becomes:

$$F = mg - \frac{\rho_f}{\rho_s}mg - C_D \pi \left(\frac{D_p}{2}\right)^2 \left(\frac{1}{2}\delta\rho_s v^2\right) \tag{7-8}$$

For a spherical particle, Equation (7-3a) = Equation (7-8),

$$m\frac{dv}{dt} = mg - \frac{\rho_f}{\rho_s}mg - C_D \pi \left(\frac{D_p}{2}\right)^2 \left(\frac{1}{2}\delta\rho_s v^2\right)$$

Divided by $m = \frac{4}{3}\pi\left(\frac{D_p}{2}\right)^3 \delta\rho_s$ on both sides, then

$$\frac{dv}{dt} = \left(\frac{\rho_s - \rho_f}{\rho_s}\right)g - \frac{3}{4}C_D\frac{1}{D_p}v^2 \tag{7-9}$$

The terminal velocity $v = v_t$ at $\frac{dv}{dt} = 0$

$$0 = \left(\frac{\rho_s - \rho_f}{\rho_s}\right) g - \frac{3}{4} C_D \frac{1}{D_p} v^2$$

$$v_t = \left[\frac{4}{3}\frac{D_p}{C_D}\left(\frac{\rho_s - \rho_f}{\rho_s}\right) g\right]^{0.5} \quad \text{for Newton's law} \tag{7-10}$$

where

C_D is the drag coefficient or coefficient of resistance. Dimensionless, $C_D \approx 0.44$.

If the fluid is water, the terminal velocity is

$$v_t = k_n \left[D_p\left(\frac{\rho_s - 1}{\rho_s}\right)\right]^{0.5} \quad \text{for water } \rho_f = 1$$

where $k_n = 1.74\sqrt{g}$

7.1.3 Allen Equation for Intermediate Flow Region

The terminal velocity of particles at an intermediate flow region was derived by Allen as shown below:

$$v_t = \left[C_1\left(\frac{\rho_s - \rho_f}{\rho_s}\right) g\right]^{2/3} \frac{D_p - hD'_p}{2\left(\frac{\mu}{\delta}\right)^{1/3}}$$

$$\cong k_A \left[D_p \frac{(\rho_s - \rho_f)}{\rho_s} g\right]^{2/3} \tag{7-11}$$

where C_1 and h are constants, D_p is the largest particle size.

7.2 Equal Settling Terminal Velocity

7.2.1 Free Settling

The diameter ratio of two particles with different densities but the same terminal velocity ($v_{tL} = v_{tH}$) for free settling condition for Stokes' criteria is

$$v_{tL} = \frac{D_{pL}^2 (\rho_{sL} - \rho_f)\delta\, g}{18\,\mu} = v_{tH} = \frac{D_{pH}^2 (\rho_{sH} - \rho_f)\delta\, g}{18\,\mu} \tag{7-12a}$$

$$D_{pL}^2(\rho_{sL}-\rho_f)=D_{pH}^2(\rho_{sH}-\rho_f) \tag{7-12b}$$

The diameter ratio of light particles (L) to heavy particles (H) at the same settling velocity in the free settling condition is

$$\frac{D_{pL}}{D_{pH}}=\left[\frac{(\rho_{sH}-\rho_f)}{(\rho_{sL}-\rho_f)}\right]^{1.0} \quad \text{for Stokes' criteria} \tag{7-13a}$$

$$\frac{D_{pL}}{D_{pH}}=\left[\frac{(\rho_{sH}-\rho_f)}{(\rho_{sL}-\rho_f)}\right]^{2/3} \quad \text{for Allen Equation} \tag{7-13b}$$

$$\frac{D_{pL}}{D_{pH}}=\left[\frac{(\rho_{sH}-\rho_f)}{(\rho_{sL}-\rho_f)}\right]^{0.5} \quad \text{for Newton's criteria} \tag{7-13c}$$

$$\frac{D_{pL}}{D_{pH}}=\left[\frac{(\rho_{sH}-\rho_f)}{(\rho_{sL}-\rho_f)}\right]^{0.5\leqslant n\leqslant 1.0} \quad \text{for general form} \tag{7-13d}$$

where L denotes lighter particles and H denotes heavier particles.

7.2.2 Hindered-Settling

In an upward flow field environment, if the falling velocity of the particles equals to the velocity of upward flow, the suspension particles will not float upward or sink downward. If the suspended particles are exceedingly crowded, the specific gravity or density of the mixture of water and particles becomes greater than unity. If another material is added to the aqueous solution, the falling velocity of the particles will be retarded by the suspended medium and reduced. The settling diameter ratio under this circumstance is called the hindered-settling ratio.

For a hindered-settling condition, the fluid density ρ_f is replaced by the apparent density of the suspension, ρ_m, to obtain a generalized equation for the hindered-settling ratio.

For two particles that have different densities, but have the same terminal velocity ($v_{tL}=v_{tH}$), the ratio of their diameters for hindered-settling condition is in a general form,

$$\text{Concentration ratio}=\frac{D_{pL}}{D_{pH}}=\left[\frac{(\rho_{sH}-\rho_m)}{(\rho_{sL}-\rho_m)}\right]^{0.5\leqslant n\leqslant 1.0} \tag{7-14}$$

where

n = 1.0 for Stokes' criteria

n = 2/3 for Allen equation

n = 0.5 for Newton's criteria

ρ_m = is specific gravity of the suspension medium

7.3 Concentration Criteria

The applicability of gravity concentration to separate coal from mineral matter can be obtained by making use of the particle ratio as the concentration criterion within particle pairs of different specific gravity and approximate particle sizes. The criterion is used to predict the effectiveness of any gravity concentration processes (Table 7.1).

Table 7.1 Concentration criterion

Concentration Criterion	Size range of applications
> 2.5	Separation easy down to fines (US Series No.200, 75μm)
2.5~1.75	Separation easy down to fines (US series No.100, 150μm)
1.75~1.50	Separation possible to 2mm (US Sieve No.10, 2mm)
1.50~1.25	Separation possible to 6.35 mm, but difficult

Example 1: Concentration Ratio for Free and Hindered Settling – Stokes' criteria

When two particles have different densities, but have the same terminal velocities, the diameter of the light particle, i.e., coal, will be greater than the heavy particle. For example, under the theory of viscosity resistance condition (Stokes' criteria):

Concentration ratio for free settling is:

$$\frac{D_{pL}}{D_{pH}} = \left[\frac{(2.65-1.00)}{(1.50-1.00)}\right]^{1.0} = \left(\frac{1.65}{0.5}\right)^{1.0} = 3.3$$

$$D_{pL} = 3.3\, D_{pH}$$

Concentration ratio for hindered settling:

$$\frac{D_{pL}}{D_{pH}} = \left[\frac{(2.65-1.30)}{(1.50-1.30)}\right]^{1.0} = \left(\frac{1.35}{0.2}\right)^{1.0} = 6.75$$

$$D_{pL} = 6.75\, D_{pH}$$

Example 2: Concentration Ratio for Free and Hindered Settling – Newton's criteria

Two particles have different densities, if they have the same terminal velocities, the diameter of the light particle, i.e., coal, will be greater than the heavy particle. For example, under the theory of turbulent flow conditions (Newton's criteria):

Concentration ratio for free settling:

$$\frac{D_{pL}}{D_{pH}}=\left[\frac{(2.65-1.00)}{(1.50-1.00)}\right]^{0.5}=\left(\frac{1.65}{0.5}\right)^{0.5}=1.82$$

$$D_{pL}=1.82\ D_{pH}$$

Concentration ratio for hindered settling:

$$\frac{D_{pL}}{D_{pH}}=\left[\frac{(2.65-1.30)}{(1.50-1.30)}\right]^{0.5}=\left(\frac{1.35}{0.2}\right)^{0.5}=2.60$$

$$D_{pL}=2.60\ D_{pH}$$

Concentration ratio is used to predict the effectiveness of any gravity concentration process. From the results of examples 1 and 2 above, it can be concluded as follows:

① The greater the value of the D_{pL} to D_{pH} ratio, the easier is the separation. This means that two particles that have larger specific gravity differences are easier to separate.

② As the ratio of D_{pL} to D_{pH} increases, the specific gravity of dense-medium increases.

③ As the particle size becomes finer, the ratio of D_{pL} to D_{pH} decreases (see Example 1 in low Reynolds flow region, and Example 2 in high Reynolds flow region).

④ The larger the ratio of D_{pL} to D_{pH}, the wider the size range can be separated.

⑤ In hindered-settling conditions, the ratio of D_{pL} to D_{pH} is greater than that in free-settling condition. This means a wider range of particle sizes can be separated in the hindered-settling condition than in the free-settling condition.

⑥ As the specific gravity of the dense-medium, ρ_m, increases, the ratio of D_{pL} to D_{pH} increases, the separation by specific gravity differences in clear layers becomes more efficient (i.e., separation becomes easier).

7.4 Separation by Difference in Settling Rate

Figure 7.2 illustrates the settling velocity of coal and quartz particles as a function of time. There are four different factors that affect solid particle behaviors in fluid:

① A coal particle (lighter particle) of 3.3 times (Stokes' criteria region) or 1.82 times (Newton's criteria region) as large as a quartz particle (heavier particle) settles to same terminal velocity. The diameter of coal particle must be smaller than that of quartz particle to avoid settling faster than the quartz particle.

② For two particles of the same specific gravity, but having different particle size, the larger size will settle faster due to a greater initial settling rate, (dv/dt) when $t = 0$.

③ For two particles of the same size, but having different specific gravities, the heavier particle will settle faster than the lighter one.

④ For two particles of the same size and the same specific gravity, the settling velocity depends on the shape of the particles on the order of spherical > cube > elongated > disc.

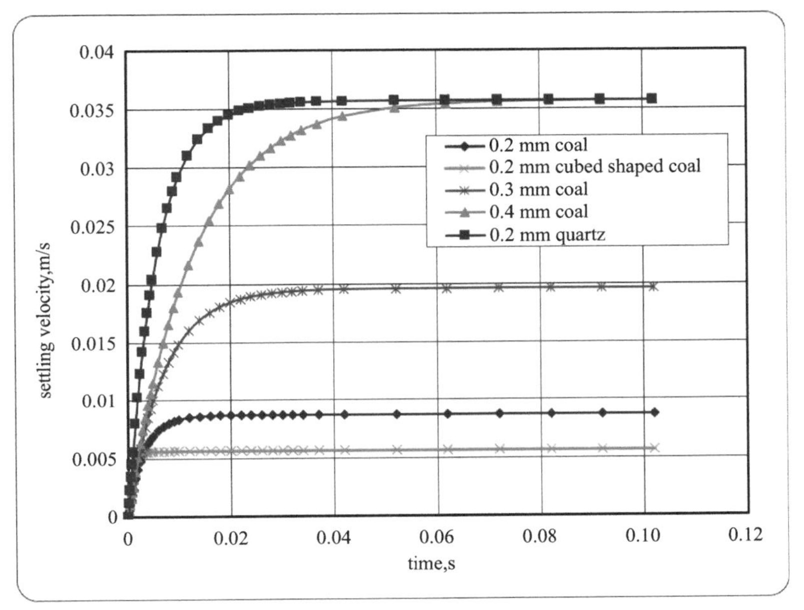

Figure 7.2 Settling velocity of coal and quartz particles as a function of time

7.5 Sphericity of a Particle

The shape of a particle can be characterized by either using a sphericity, φ, or shape factor, λ = 1/φ, that denotes the deviation in shape of a particle from a spherical shape. The ratio of the surface area to volume of a sphere with diameter D_p, is the specific surface area of the sphere, So. the specific surface area of a spherical particle, S_{os}

$$S_{os}=\frac{\text{Surface area of a sphere}}{\text{Volume of a sphere}}=\frac{4\pi\left(D_p/2\right)^2}{4/3\,\pi\left(D_p/2\right)^3}=\frac{6}{D_p} \tag{7-15a}$$

Specific surface area of an actual particle, S_{oa}

$$S_{oa}=\frac{\text{Surface area of an actual partcle}}{\text{Volume of an actual particle}} \tag{7-15b}$$

Sphericity, φ is defined as

$$\varphi=\frac{S_{os}}{S_{oa}}=\frac{\text{Specific surface area of a shere having an equivalent volume of actual particle}}{\text{Specific surface area of actual particle}}$$

Example 1: An actual particle has a cubic shape as shown in Figure 7.3.

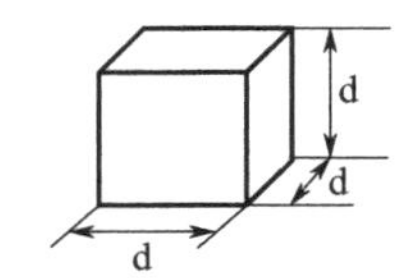

Figure 7.3 Cubic shaped particle

Determine the sphericity of the particle.

The volume of an actual cubical particle with width d = d^3

The specific surface area of the actual cubical particle is $S_{oa} = 6d^2/d^3 = 6/d$

Volume of a sphere = $4/3\ \pi\ (D_p/2)^3$

Volume of a sphere = Volume of the cube:

$$\frac{4}{3}\pi\left(\frac{D_p}{2}\right)^3 = d^3$$

Solve for D_p

$$D_p = d\left(\frac{6}{\pi}\right)^{1/3}$$

Sphericity of a cube:

$$\varphi = \frac{S_{os}}{S_{oa}} = \frac{\frac{6}{D_p}}{\frac{6}{d}} = \frac{d}{D_p} = \frac{d}{d\left(\frac{6}{\pi}\right)^{1/3}} = \left(\frac{6}{\pi}\right)^{1/3} = 0.806$$

Shape factor: $\lambda = 1/\varphi = 1/0.806 = 1.24$. Table 7.2 lists the shape factors of various materials related to coal cleaning.

Table 7.2 Sphericity and shape factor for various screened materials ($\lambda = 1/\varphi$)

Material	Sphericity, φ	Shape factor, λ
Sand, jagged	0.595	1.68
Sand, nearly spherical	0.870	1.15
Sand, angular	0.671	1.49
Sand, flakes	0.394	2.54
Sand, round	0.807	1.24
Crushed glass	0.650	1.54
Coal, pulverized	0.730	1.37
Coal, dust up to 3/8-in	0.650	1.54
Coal	0.581	1.72
Crushed and screen ores	0.571	1.75
Limestone	0.455	2.20
Flake graphite	0.126	7.96

7.6 Unit Operations for Coal Concentration

Figures 7.4 and 7.5 show the typical efficient separating gravity range of commercial gravity concentrators and preferred size ranges of feeds to major coal cleaning devices, respectively.

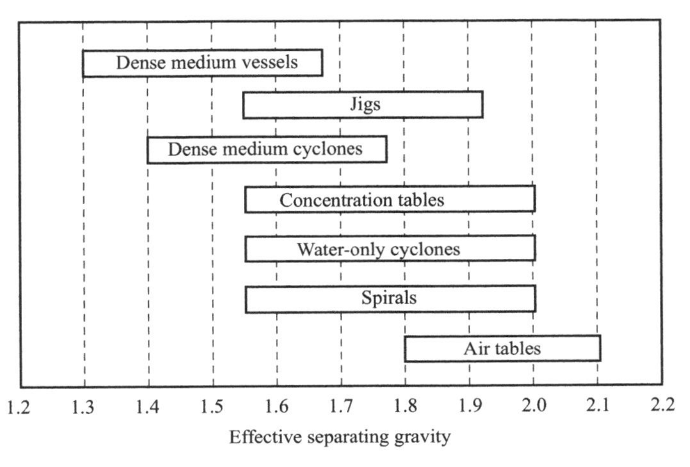

Figure 7.4 Typical efficient separating gravity range of commercial gravity concentrators

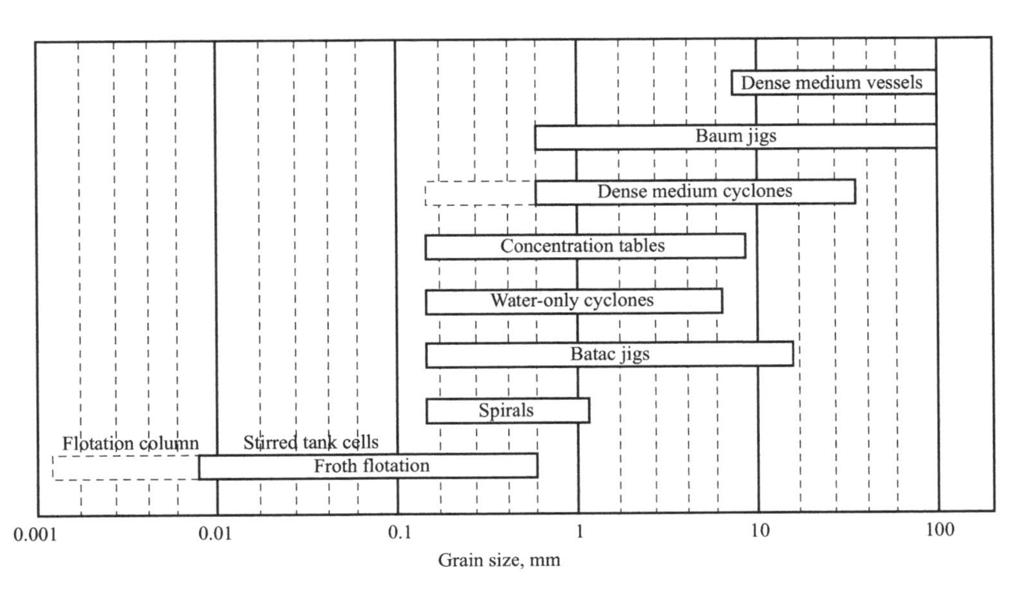

Figure 7.5 Preferred size ranges of feeds to major coal cleaning devices

7.6.1 Jig

In jigging, a mixture of particles is supported on a perforated plate on screen in a layer or "bed" with a depth many times the thickness of the largest particles. The mixture is subjected to an alternating rising and falling (pulsating) flow of fluid. The objective is to cause all the particles of high specific gravity to travel to the bottom of the bed, while the particles of lower specific gravity collect at the top of the bed. The fluid can be liquid or gas. The typical jig used in a coal preparation plant is shown in Figure 7.6.

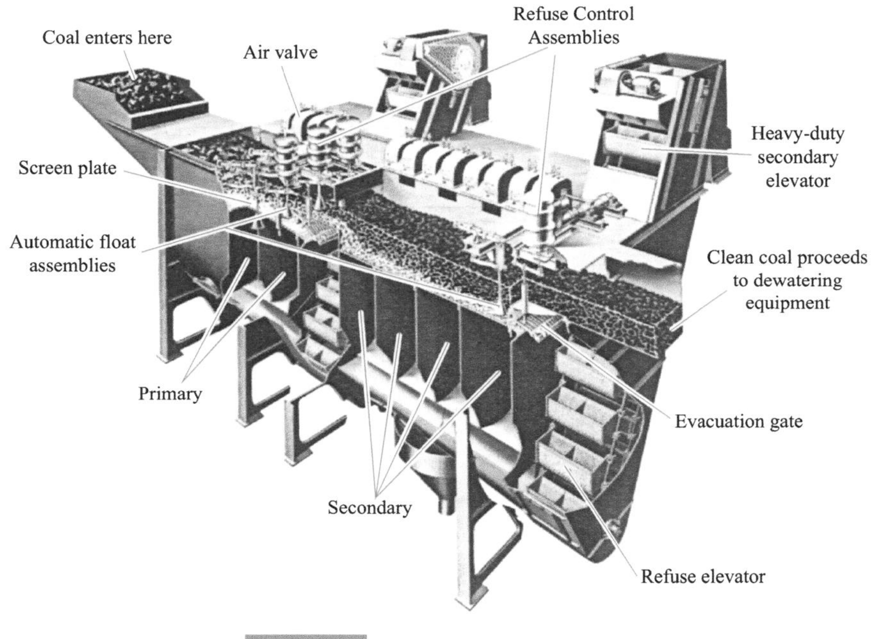

Figure 7.6 McNally Mogul washer

Jigging is commonly used for the separation of coarse coal. (The jig can also be designed to process fine coal). It is relatively cheap in construction, operation and maintenance, and is relatively unaffected by the grade of feed. Figure 7.7 is a diagram of jig operation. The general principle of the jigging operation is the principle of particle motion under pulsed-flow and hindered-settling conditions. The separating mechanisms are: ① Hindered settling, ② Pulse flow acceleration, and ③ Dense-bed penetration and percolation.

The liquid pulse is essentially a modulated sine wave. Inside the separation chamber, a layer of heavy particles at the bottom of the bed controls the rate at which the fine concentrate particles penetrate through the bed to the hutch. With some ore or coal, there is enough of coarse heavier minerals to provide this layer. With coals or ore, it is necessary to provide an added layer of coarse heavy mineral which is called "ragging". Generally, the ragging particles must be heavy enough to remain at the very bottom of the bed, but light enough for dilation in the up-stroke. The size must be greater than the screen openings, and large enough to provide spaces between the particles to allow concentrate particles to percolate through on the down stroke. In coal jigging, feldspar ragging may be used. The feldspar chips, with the thickness to length ratio of 1 to 3, teeter on edge during pulsation stroke and act as valves to allow refuse pieces to pass through the screen, while closing off the passage to coal particles and to modulate the suction part of the pulse wave. Shale pieces from the jig bed can be used for the same effect.

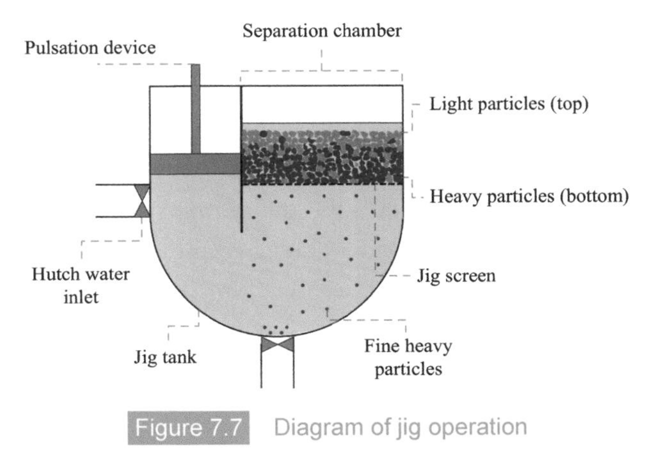

Figure 7.7 Diagram of jig operation

The capacity of the Baum jig is usually 3-6 TPH per sq. ft. of active screen area. The machine capacity ranges from 25 to 100 TPH. Water requirement is usually 1000 to 2000 gpm. The velocity of the strokes is 6-in per second, and the number of pulsations is 6 per minute.

$$V = nA\pi/60$$

$$6\ \text{in/sec} = (6\ \text{pulse/min})A\pi/60$$

Thus $\quad A = 2$-in stroke length or amplitude.

The power requirement is approximately 0.1 HP per sq. ft. of screen area, and compressed air supply need is 3 psi. Operating controls are necessary to control bed density and level, usually by varying the rate of product removal. The sensors are usually of the balanced-float-compartment type. Submerged stream-lined floats in bed to sense density and automatically control either an entrance or exit to modify the jig stroke may also be used.

(1) Hindered settling

① For two particles of the same size $D_{PL} = D_{PH}$ but different specific gravity, the heavier particle has a greater terminal velocity while the lighter particle has a smaller terminal velocity. For example, for a coal particle of 1.5 specific gravity and a mineral such as quartz particles of 2.7 specific gravity in medium suspension of 1.2 specific gravity, the hindered settling velocity ratio under Stokes condition is

$$\frac{(v_t)_L}{(v_t)_H}=\left[\frac{(\rho_L-\rho_m)/\rho_m}{(\rho_H-\rho_m)/\rho_m}\right]^{0.5-1.0}=\left[\frac{(1.5-1.3)/1.3}{(2.7-1.3)/1.3}\right]^{0.5-1.0}=\left[\frac{1}{7}\right]^{0.5-1.0}$$

Therefore, the terminal velocity of the quartz particle $(v_t)_H$ is 2.65 to 7 times that of the same size coal particle $(v_t)_L$.

② If the coal particle and quartz particle have the same terminal velocity, $(v_t)_L=(v_t)_H$, that means the heavier particle has a smaller diameter and the lighter particle has a larger diameter.

$$\frac{(D_p)_L}{(D_p)_H}=\left[\frac{(\rho_L-\rho_m)}{(\rho_H-\rho_m)}\right]^{0.5-1.0}=\left[\frac{(1.5-1.3)}{(2.7-1.3)}\right]^{0.5-1.0}=\left[\frac{1}{7}\right]^{0.5-1.0}$$

Therefore, for the coal particle that has the same terminal velocity as the quartz particle, the size of the coal particle must be 2.65 to 7 times greater than that of the quartz particle as shown in Figure 7.8(a).

(2) Differential initial acceleration

The coal particle and quartz particle have different specific gravities, thus, the initial accelerations are different. The initial velocity is the acceleration velocity of the particle from the static state. From Equation (7-4)

$$\frac{dv}{dt}=\frac{\rho_s-\rho_m}{\rho_m}g-\frac{1}{2}\frac{\rho_m v}{\left(\frac{D_p}{2}\right)^2\rho_s}$$

At initial state,

$$\frac{1}{2}\frac{\rho_m v}{\left(\frac{D_p}{2}\right)^2\rho_s}=0$$

Therefore,

$$\left(\frac{dv}{dt}\right)_{t=0}=\frac{\rho_s-\rho_m}{\rho_m}g$$

The initial velocity of a coal particle:

$$\left(\frac{dv}{dt}\right)_{t=0}=\frac{1.5-1.3}{1.3}g=0.154g$$

The initial velocity of a quartz particle:

$$\left(\frac{dv}{dt}\right)_{t=0}=\frac{2.7-1.3}{1.3}g=1.077g$$

$$\frac{\left(\frac{dv}{dt}\right)_{t=0,L}}{\left(\frac{dv}{dt}\right)_{t=0,H}}=\frac{0.154}{1.077}=\frac{1}{7}$$

That means the quartz particle has 7 times more the initial acceleration than the coal particle. The initial slopes for the coal particle and quartz are different. The initial acceleration of the quartz particle has a steeper slope than that of the coal particle. The difference in the initial velocity of the two particles is dependent upon the specific gravity of the particles, but is independent of their particle size. Thus, disregarding the size of coal and quartz particles, the quartz particles will always have larger initial acceleration than the coal particles. In jigging operations, the period is short, and the frequency is high. The jigging process applied this principle of different initial acceleration for different particle sizes. Also, the hindered settling becomes more significant due to the differences in the initial acceleration between particles as shown in Figure 7.8(b).

(3) Consolidation trickling

In jigging operations, when the cycle is at the static state or suction stage, the settled particles accumulate on the surface of the sieve. Larger particles cannot settle close to the sieve due to other particles preventing them from settling further. However, the smaller particles may pass through the interstitial of the particle bed and continue to fall. This will result in the smaller particles settling further than the larger size particles in the bed as shown in Figure 7.8(c).

(4) sorting

In the jigging process, due to the alteration of upward current and static stage or suction stage with three major phenomena occurring in the process as described in the previous sections, it results in the heavier particles arranging in the lower part of the bed. Thus, the coal can be separated from the minerals as shown in Figure 7.8(d).

An optimum jig operation requires: ① maximizing density stratification by properly lifting, opening and closing the bed on each cycle, ② providing the maximum number of jigging cycles before discharge, ③ monitoring the refuse layer in each compartment, and ④ controlling the rate of refuse removal from the bed. The control can be programmed for constant operation or to utilize the patented regulation principle to vary the stroke automatically as the jig float senses a change in feed conditions, and the system will provide superior separation results. Figure 7.9 explains the operation of the control system.

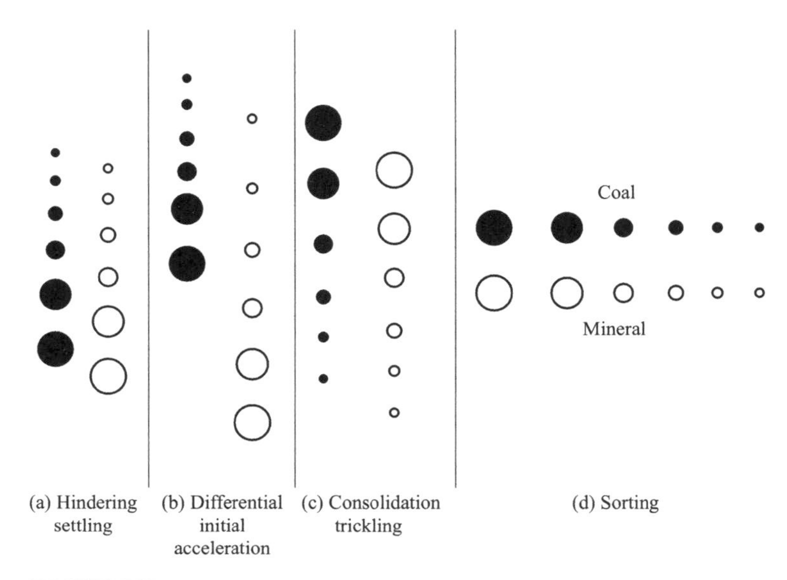

Figure 7.8 Principle of coal and mineral particles moving in jigging

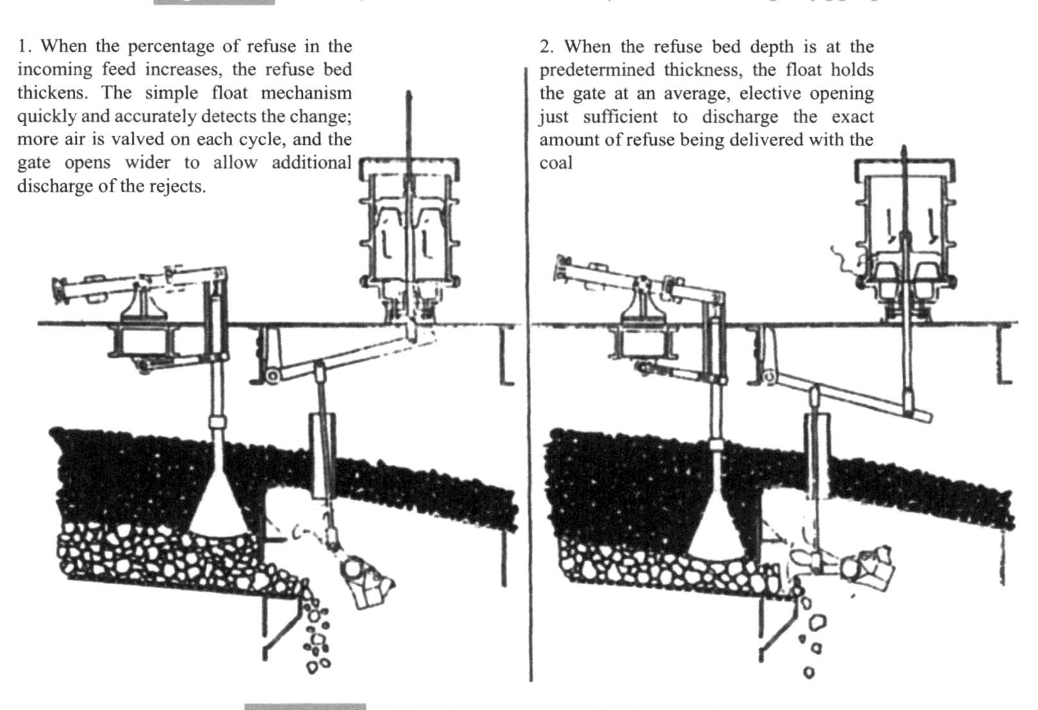

Figure 7.9 How the automatic float control works

7.6.2 Dense Medium Vessel

Dense medium vessel utilizes the suspension of magnetite to adjust the medium density to a value that is between coal and ash bearing material to separate coal from refuses. The medium density correlates to the required separation density needed to achieve a desired product grade. The feed to a heavy media vessel is injected perpendicular to the elongated width of the vessel. The float material travels across the vessel or is removed by a flight conveyor at one end as the coal product. The high density particles sink to the moving chain conveyor to other end.

There are several types of dense medium vessels as shown in Figure 7.10:

① Single-gravity, two product system with a circular weir which can process large particle sizes.

② Single-gravity, two product system with a rectangular weir which provides large float capacity classified feed.

③ Dual-gravity, three product system with optional weir sections.

④ Dual-gravity, four product system with independent media circuits and optional weir sections.

7.6.3 Dense Medium Cyclone

Dense medium cyclone separators have now been widely used for processing different ores and coal. They provide a high centrifugal force and low viscosity in medium and can process much finer size coal feed. Research has shown that good separation could be obtained for coal particles as fine as 0.1 mm. The principle of dense medium cyclone operation is similar to that of the conventional hydrocyclone. The coal is suspended in the medium and introduced tangentially to the cyclone by pump or gravity. As shown in Figure 7.11, the refuse as heavier material is centrifuged to the cyclone wall and exits at the apex and the light product goes to the flow around the axis and exits via the vortex finder. Figure 7.12 shows the process for application of dense medium cyclone for coal preparation.

7.6.4 Concentration Table

Concentration table utilizes the horizontal current to achieve material separation. It is the result of the different velocity of a flowing liquid film, being maximum at or near the top of the film and almost nil at the bottom. Flowing film concentration results in the coarse low density particles being entrained in the upper flowing film layer. Consequently, these particles are moving at maximum velocity. Conversely, the fine high density particles report

General specification data

Single-gravity systems

Single drum
Sizes: 6′to 12′diameter
Capacities: to 400 TPH
Power requirements 5 to 20 HP
Single drum, dual overflow
Sizes: 6′x11′to 15′x23′diameter
Capacities: to 900 TPH
Power requirements: 10 to 30 HP

Dual-gravity systems

Dual-gravity, three-product drum
Sizes: 6′x10′to 12′x20′
Capacities: to 325 TPH
Power requirements: 10 to 20 HP
Dual-gravity, four-product drum
Sizes: 6′x10′to 12′x20′
Capacities: to 650 TPH
Power requirements: 10 to 20 HP

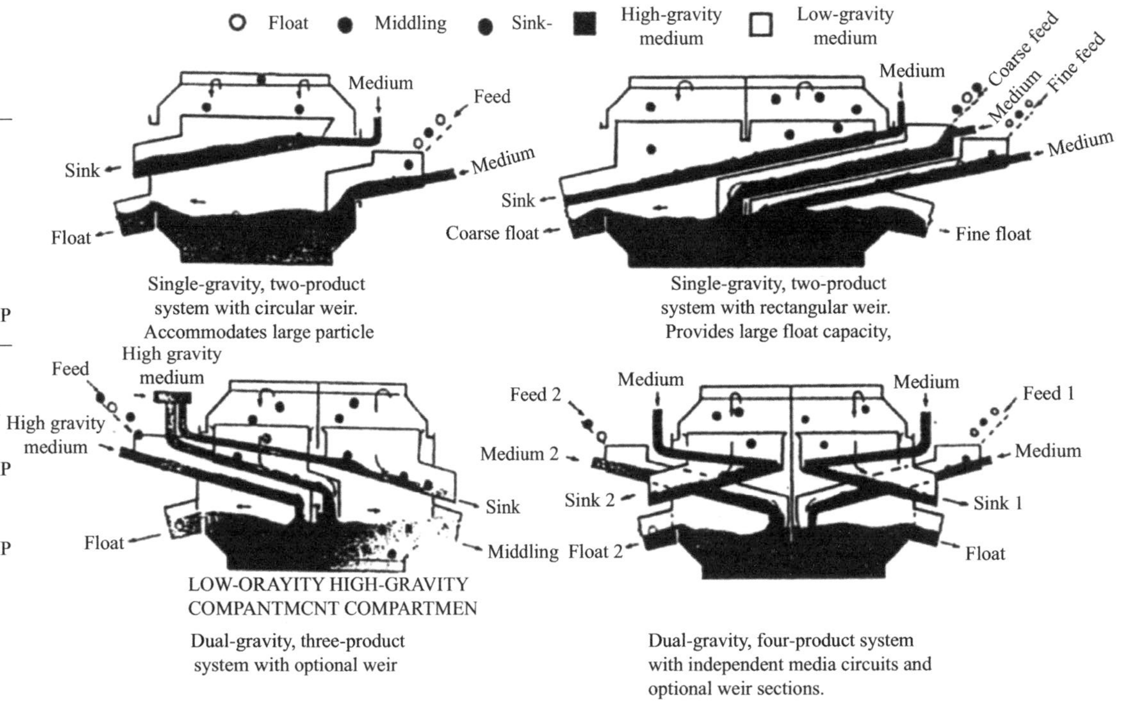
Float
Middling
Sink-
High-gravity medium
Low-gravity medium
Medium
Feed
Sink
Medium
Float
Single-gravity, two-product system with circular weir. Accommodates large particle
Medium
Coarse feed
Medium
Fine feed
Medium
Sink
Coarse float
Fine float
Single-gravity, two-product system with rectangular weir. Provides large float capacity,
Feed
High gravity medium
High gravity medium
Sink
Float
Middling
LOW-ORAYITY HIGH-GRAVITY
COMPANTMCNT COMPARTMEN
Dual-gravity, three-product system with optional weir
Feed 2
Medium
Medium
Feed 1
Medium 2
Medium
Sink 2
Sink 1
Float 2
Float
Dual-gravity, four-product system with independent media circuits and optional weir sections.

Figure 7.10 Diagram of drum vessel separation system

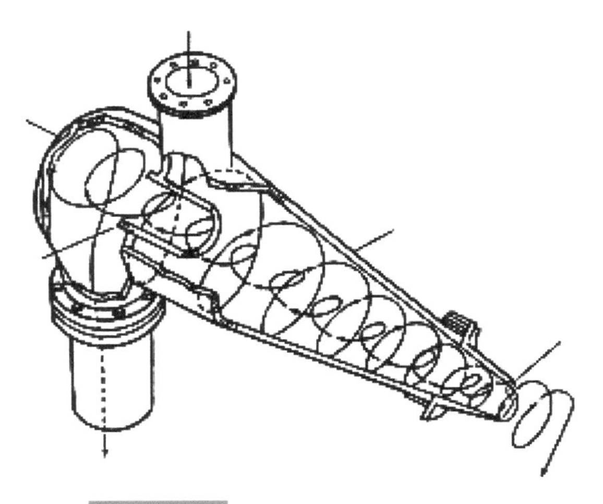

Figure 7.11 Dense medium cyclone

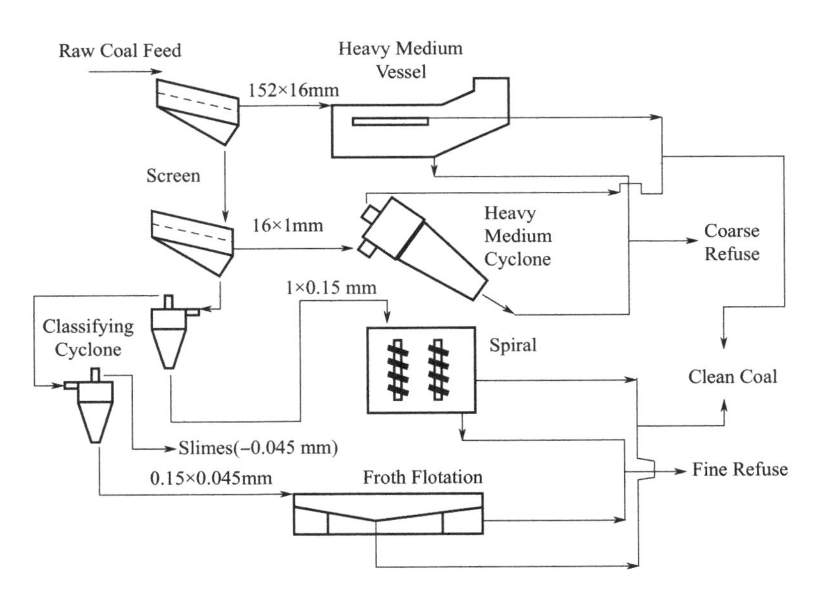

Figure 7.12 Process for application of dense medium cyclone for coal preparation

to the bottom of the flowing film layer and move at minimum velocity. When two particles of the same size but of different specific gravity are considered, the higher density one moves more slowly by reason of its greater mass. As a result, the particles tend to become arranged in the manner shown in Figure 7.13. The velocity of the pulp is greatly increased by the presence of new colloidal particles and consequently, the penetration of the somewhat larger grains through the flowing film is retarded. It results in them being carried further across the concentrator deck. Obviously, this effect can be reduced by prior removal of some or all of the ultrafine particles. The detrimental effect of kaolin on the separation has been reported. The pH value also has a very remarkable influence on the viscosity. The use of acid mine water for instance can increase viscosity materially and thereby results in lower efficiency.

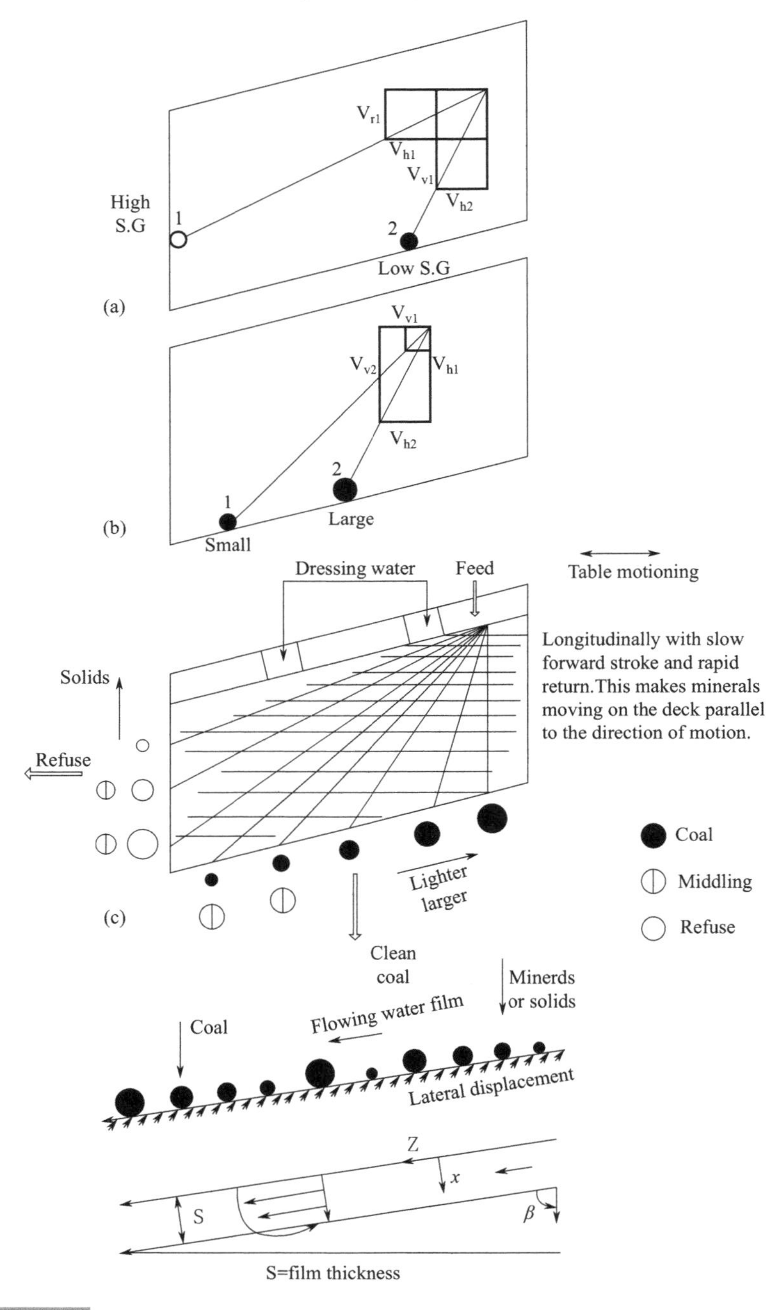

Figure 7.13 The motion of minerals and coal particles on the concentration table

(1) Hindered settling

This is a specific gravity sorting process that occurs on the table when the densely packed suspension concentrate between the riffle is pulsed. The resulting classification of particles is similar to that secured by flowing film concentration.

(2) Consolidation trickling

This phenomenon occurs subsequent to hindered settling in which fine particles continue to settle into the voids remaining after the coarse particles have settled.

(3) Asymmetrical acceleration

This is the result of "reciprocating asymmetrical motion" imparted to the deck which causes intermittent travel of solids resting on the deck.

In the operation, a slurry of solids and water is fed to the upper edge of the sloping table. As the suspended material moves across the table, it is caught and forms pools behind the longitudinal riffles. The differential shaking action of the deck causes size classification and specific gravity stratification. The result is that particles with similar specific gravity become arranged vertically according to size.

Once the coal bed is formed on the deck, the addition of moving slurry and action of the flow of cross water causes shearing of the top layers of the stratified particles. Thus, it forces the lower specific gravity and coarser particles to cascade over the riffles toward the lower side of the table. The depth of the riffles and the bed thickness decrease from the drive mechanism end to the discharge end of the table. This results in continuous flowing film concentration of increasing finer sizes and higher density particles as these particles move longitudinally along the table.

(4) Performance of concentration table (Table 7.3 and 7.4)

Table 7.3 Capacity of concentration table

Top size of feed	-3/4 inch	-3/8 inch	-28 mesh
TPH/deck	15	12.5	5

Speed and stroke are 250~300 rpm (stroke per minute) for ¾ inch to 3/8 inch size, respectively.

Table 7.4 Water consumption

Top size of feed, inch	Water to coal ratio	Remark
3/4	2.5~3 : 1	Additional water to wash out refuse from the bed
7/16	2 : 1	
3/16	2 : 1	
28 mesh	1.5~3 : 1	To reduce the viscosity of feed slurry

Power required is 1-3 HP per deck for single deck; 3 HP for double and triple decks.

The differential drive mechanism is designed so that at the end of backward stroke (deck moving toward the drive end), the deck and hence the particles on the deck surface are momentarily at rest. The deck is then accelerated forward until at the end of stroke the direction of travel is rapidly reversed. The particles on the deck which move with the deck on the forward stroke will now slide forward due to the momentum, while the deck reserves its direction and starts on its backward stroke. Thus, the particles always travel toward the discharge end of the table. The idealized size and specific gravity stratification of the table action described here are shown in Figure 7.14. However, the height and placement of the riffles, irregularities in feed rate, deck surface, table motion, water supply and distribution all exert influences that will modify the idealized behavior of the feed pulp.

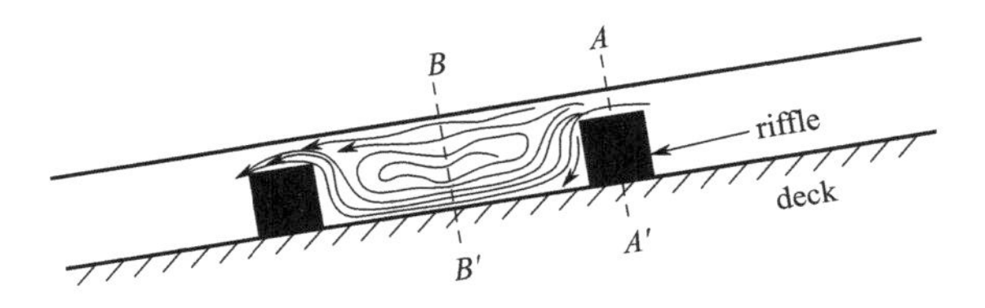

(a) Flow velocity between riffles

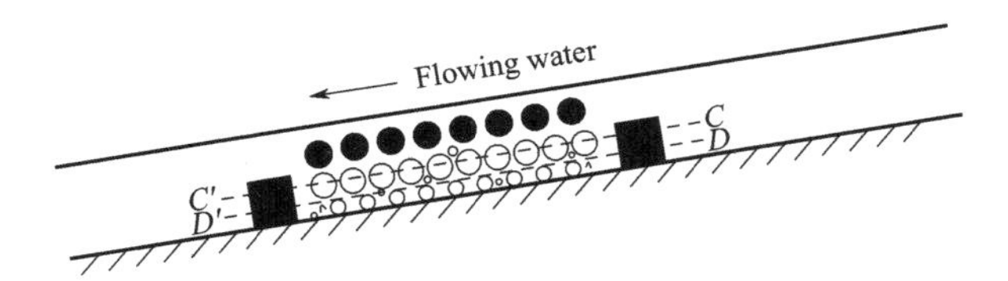

(b) Vertical stratification between riffles

Figure 7.14 Flow patterns in riffles of the concentration table

(5) Particle size and specific gravity

① The motion of particle on an un-riffled deck

i. Difference in specific gravity, but the same diameter of particles (Figure 7.13(a))

$V_{v2} > V_{v1}$, $V_{h1} > V_{h2}$

ii. Difference in diameter, but same specific gravity (Figure 7.13(b))

$V_{v2} > V_{v1}$, $V_{h2} > V_{h1}$

iii. In reality, for a given material, it consists of particles with various sizes and specific gravities. Therefore, the smallest particle with the highest specific gravity will travel toward the left end (tailing end) of the table, while the lighter larger particles travel towards the front (coal recovery end) of the table (Figure 7.13(c)).

② The motion of particles on the riffled deck

On the deck with riffles, the flow rate at cross-section AA′ is much faster than that at cross-section BB′ [Figure 7.14(a)]. Thus, the particles will accumulate between the successive riffles. Due to the motion of the table and the eddy current of the fluid, the particles between the adjacent riffles are subjected to hindered settling (Figure 7.14(a)). The larger sized particles of high specific gravity will fall to the very bottom of the particle bed, and smaller particles of high specific gravity will fall over them. Smaller sized refuse (of heavier particles) will flow following the motion of the fluid. Following the motion of the table, they travel toward the left end (tailing end) of the table. The height of riffles also gradually decreases as shown in Figure 7.14(b). At the CC′ and DD′ height of the riffle, which is pushed by the fluid flow traversely along the riffles until the particles lean in minerals reach the deck with the lowest riffle height or without riffles, and are recovered as coal rich particles.

Although the hindered settling effect along any individual riffle might be relatively slight, the cumulative effect along the entire series of riffles across the width of the deck might be of sufficient magnitude to materially influence the characteristics of the table separation.

The motion of the table is perpendicular to the water flowing on the deck. The traversely motion of the table separates the particles according to their specific gravity in a vertical motion. The water current classifies the particles according to their size at the horizontal direction. Both motions of the particles are perpendicular to each other. Thus, the motion of the particles is governed by the combined forces of these two motions and move diagonally across the table from the feed and fan out to both ends of the table (tailing and concentration ends). The order of the sorting process is that larger particles of specific gravity move to the farthest.

7.6.5 Water-Only Cyclone

A hydrocyclone utilizes centrifugal and drag forces to separate materials based on differences in physical properties such as size, density, and shape. For a particle system that contains a mixture of two different-density materials, the separation between two components can be achieved by concentrating the mixture into dense and light fractions. In classifying hydrocyclone separations, density and size effects occur simultaneously, with the size effect dominating. Hence, the separation of the particles into light and heavy fractions are not possible. In order to increase the density effect, it is necessary to modify the geometry of the hydrocyclone. The modified hydrocyclone is termed a water-only cyclone. Usually, the water-only cyclone has a wider cone angle and a longer vortex finder (Figures 7.15 and 7.16) as compared to a conventional classified cyclone. By increasing the cone angle, the effect of particle density can be enhanced. The wide-angle results in a build-up of an autogenous bed of particles along the conical section of the hydrocyclone. Its bed prevents particles from moving unimpeded, and "hindered-settling" conditions prevail.

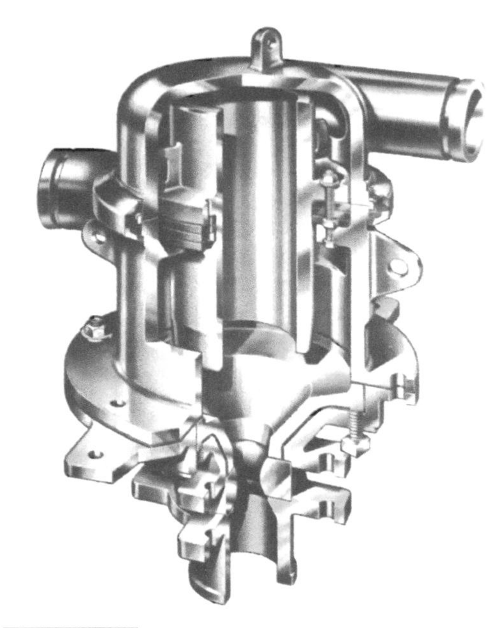

Figure 7.15 Cross section view of the McNally Visman Triclone

Another important modification is the longer vortex finder of a water only cyclone. Extending the vortex finder toward the underflow opening reduces the distance between the end of the vortex finder and the wall of the conical section of the cyclone. The relative shortness of this distance means that the vertical path traveled by particles in the upward central current of the cyclone is sufficiently reduced so that large low-

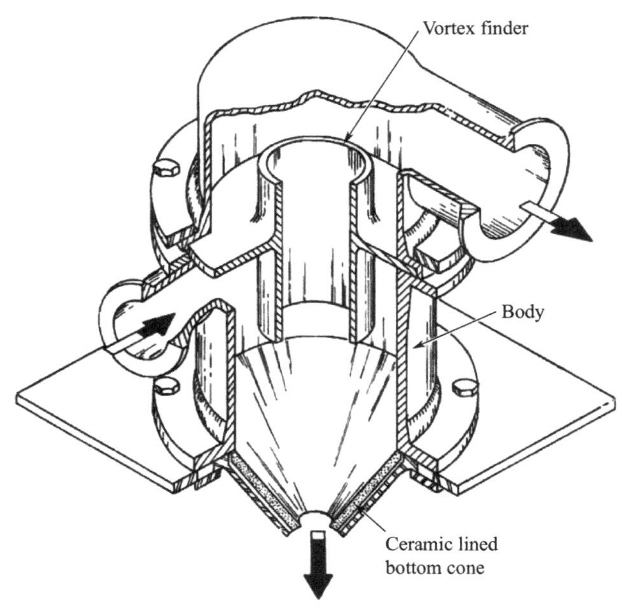

Figure 7.16 Diagram of separation sequence of McNally Visman Triclone

density particles caught in the upward current will not reach their terminal settling velocity and settle out to the wall, but will be captured inside the vortex finder and discharged with the light low-density fraction. The result of the wider cone angle and longer vortex finder is that the separation will occur primarily on the basis of differences in particle density as opposed to particle size. Water-only cyclones have been used throughout the coal and mineral industries to concentrate a high-density fraction from lower-density particles.

7.6.6 Spiral Concentrators

Spiral concentrators have found various applications in mineral processing for many years. Those designed for applications in coal preparation operations have a shallower pitch than those used in heavy mineral separation. The typical spiral unit consists of a spiral conduit with a modified semi-circular cross section. Diagrammatic view of a full size spiral concentrator is shown in Figure 7.17. As the slurry flows down the spiral, those minerals with the highest specific gravity move to the inner edge of the spiral while the light material moves upward to the outer edge of the spiral as shown in Figure 7.18.

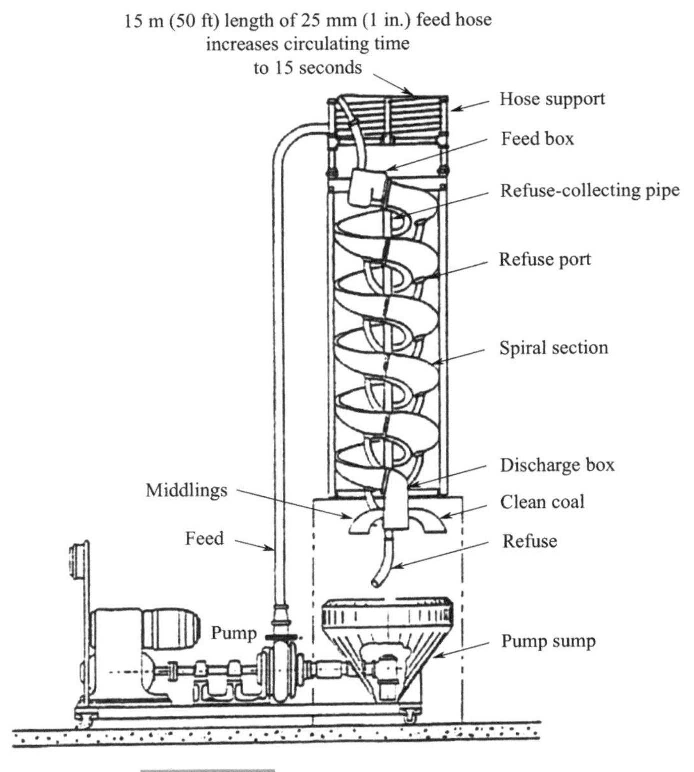

Figure 7.17 Closed circuit spiral unit

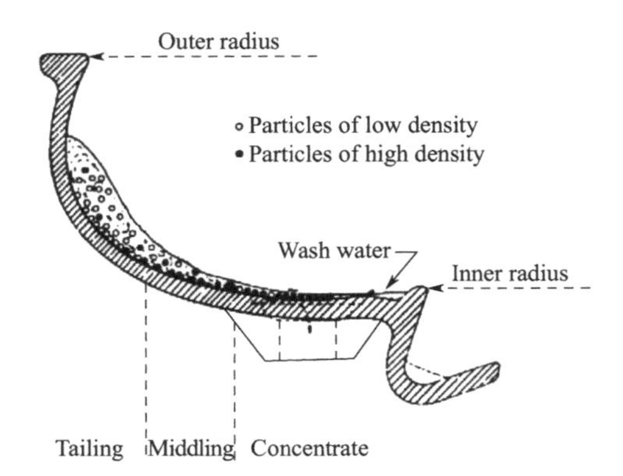

Figure 7.18 Cross section view of a spiral for coal preparation

The effectiveness of spiral separation for the removal of clean coal from pyrite and other contaminants varies according to the type of coal feed and the feed rate. For coals containing 19.0% to 41.9% ash the spiral separator can produce products containing less than 9% ash and up to 98.7% coal. As the feed rate ranges from 1.1 TPH to 3.9 TPH per spiral start, coal recovery is maintained at levels of 90% over the full range of particle sizes.

The shallow pitch spirals used for the beneficiation of coal have capacities in the range of 1 to 3 TPH per spiral start. A very recent study has suggested that while spirals will accept up to 3.5 TPH per spiral start, optimum performance is achieved at a feed rate of 2 TPH per spiral start. However, ash displaced to the product increases with increasing feed rate and particle sizes below 0.10mm. It has also been stated that feed rate is in fact the most significant independent variable governing spiral performance. Table 7.5 shows the spiral separation results on Ohio refuse slurry (-6 mesh).

Table 7.5 Spiral separation results on Ohio refuse slurry (6×0 mesh) (Coal product dewatered at 100 mesh)

Product	kg/h	wt%	Analysis			Distribution, %		
			Ash %	S %	kj*	Ash	S	kj*
Set No. 1								
Calculated feed	1567	100.0	27.81	7.46	10535	100.0	100.0	100.0
Clean coal	938	59.9	8.91	3.61	13518	19.2	29.0	78.4
−100 mesh	53	3.4	63.37	3.69	4604	7.5	1.6	1.5
Refuse	576	36.7	55.39	14.06	5637	73.3	69.4	20.1
Set No. 2								
Calculated feed	1214	100.0	26.94	7.02	10529	100.0	100.0	100.0
Clean coal	726	59.8	7.62	3.28	13936	16.9	28.0	79.2
−100 mesh	70	5.8	65.31	3.67	4109	14.1	3.0	2.2

Product	kg/h	wt%	Analysis			Distribution, %		
			Ash %	S %	kj*	Ash	S	kj*
Refuse	418	34.4	54.07	14.10	5688	69.0	69.0	18.6
Circulated middling	304		16.59	4.66	12281			
Set No. 3								
Calculated feed	1239	100.0	25.26	7.07	10684	100.0	100.0	100.0
Clean coal	776	62.6	8.00	3.18	13752	19.9	28.2	80.7
−100 mesh	49	4.0	73.37	3.91	2350	11.3	2.2	0.9
Refuse	414	33.4	52.03	14.73	5900	58.8	69.6	18.4
Circulated middling	400		18.68	4.90	1836			

* kj: kilojoule.

7.6.7 Hindered-Settling Bed Separators

Hindered-settling bed separator (Figures 7.19 and 7.20) is also called teeter-bed separator. It has been developed from the classical hindered settling classifier which is used for size classification; however, teeter-bed separator can be used to treat coal in the −3+0.25 mm size range, at cut points below 1.60. Teeter-bed separator is a vessel in which the feed settles against an evenly distributed upward current of water. When the falling feed particles achieve the same velocity as the upward current, they will not fall any longer. The feed material builds into a bed within the cell, creating a teeter zone. The lighter material overflows into the upper launder while heavier particles not supported by the upward current water settles out of the

Figure 7.19 Hindered-settling bed separators

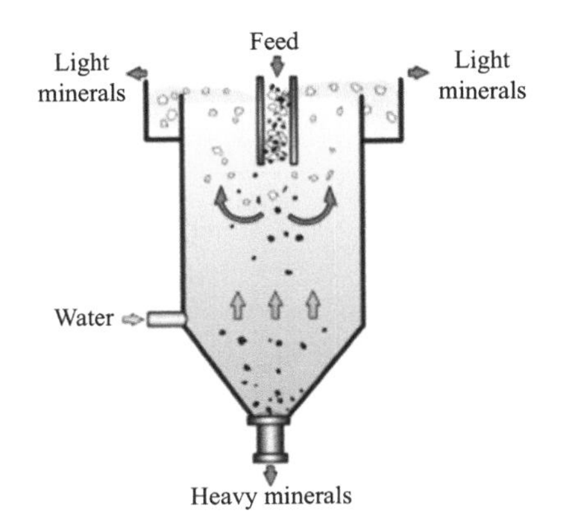

Figure 7.20 Schematic diagram of a hindered-settling bed separator

teeter zone and is discharged at the bottom of the cell. It can also be used for reprocessing of spiral products, or scavenging coal from coarse tailings. Teeter-bed technology shows advantages over the use of spirals due to their low product ash levels, lower cut points, having high mass flow capacity in the range of 20~40 TPH/m^2, low operating cost, easy layout and high levels of near-gravity material. However, it is sensitive to particle size. It can be inserted between dense medium cyclones and froth flotation.

Teeter-bed separator became quite widely used in the United Kingdom during the 1980s, particularly in coal preparation plants for cleaning up old waste piles, and eventually migrated to Australia, the United States and South Africa to treat fine coal.

7.6.8 Pneumatic Coal Preparation Process

Pneumatic coal preparation process offers some advantages and provides an efficient solution for coal cleaning by dry method. Some pneumatic equipment has been applied in dry coal cleaning processes, such as multiplex dry cleaning coal unit, fluidized dense-medium dry coal separator, allair jig, airflow jig, etc.

(1) Multiplex dry coal cleaning unit

Multiplex dry coal cleaning unit employs the separation principles of an autogenous medium and a table concentrator. Dry coal from the surge bin is fed into a separation apartment at a predetermined flowrate. The separation compartment consists of a desk, vibrator, air chambers and a hanging structure. Schematic diagram of multiplex (FGX) dry cleaning coal unit is shown in Figure 7.21. A typical flowsheet for multiplex (FGX) dry cleaning coal unit operations is presented in Figure 7.22.

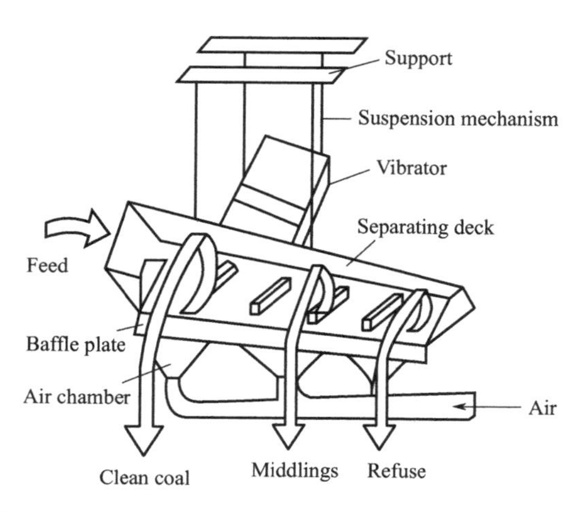

Figure 7.21 Schematic diagram of multiplex (FGX) dry cleaning coal unit

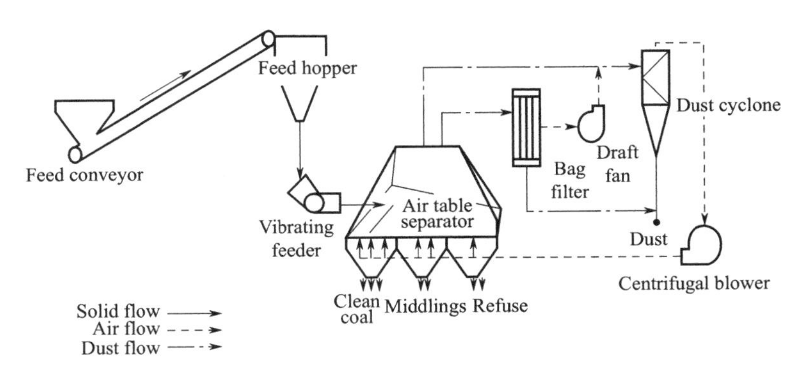

Figure 7.22 A schematic diagram of the FGX-1 dry separator test circuit

There are different types of FGX multiplex dry cleaning coal machines. They can be selected for industrial applications depending on the objectives such as feed size, capacity, etc. Type of FGX multiplex dry cleaning coal machines is listed in Table 7.6.

(2) Fluidized-bed dense-medium dry cleaning coal separator

Dry coal cleaning process utilizing fluidized-bed dense medium has drawn worldwide attention and developed rapidly in the past few years. It has the advantages of low investment cost, no water-use, high separation efficiency, wide range of separating density, etc. The main principle is that the feedstock stratifies by bed density, with the lighter particles (clean coal) floating and the denser particles (tailings) sinking based on the Archimedes theorem. Thus, the crucial solution is to maintain stable fluidization and form uniform density distribution in the bed by selecting suitable fluidized media and seeking optimal operating parameters for the effective separation of coal. The schematic diagram of the fluidized-bed dry cleaning coal separator using magnetite as dense medium is presented in Figure 7.23.

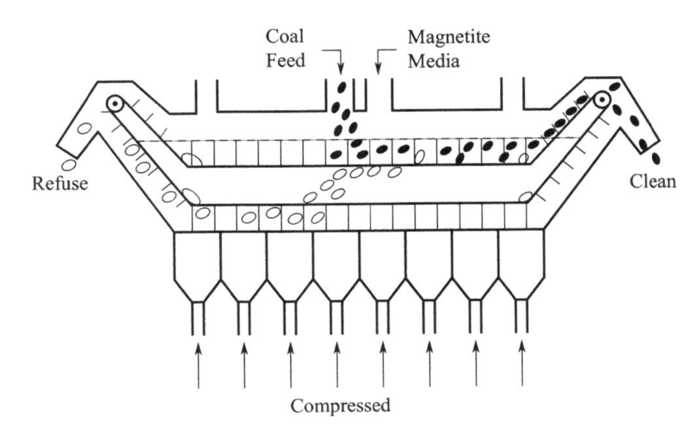

Figure 7.23 Schematic diagram of air magnetite separator

Table 7.6 Type of FGX multiplex dry cleaning coal machines

Item	Unit	FGX-1	FGX-2	FGX-3	FGX-6	FGX-9	FGX-12	FGX-18A	FGX-24A	FGX-24	FGX-48A
Area of separation		1	2	3	6	9	12	9×2	12×2	24	24×2
Feed size	mm	60-0	60-0	80-0	80-0	80-0	80-0	80-0	80-0	80-0	80-0
Surface moisture	%	<7	<7	<7	<7	<7	<7	<7	<7	<7	<7
Capacity	TPH	8-10	18-20	25-30	50-60	75-90	90-120	150-180	180-240	180-240	350-480
Efficiency	%	>90	>90	>90	>90	>90	>90	>90	>90	>90	>90
Total power	kw	24.15	59.97	73.77	142.87	248.27	323.07	448.47	640.94	635	1255
Dimension L×W×H	m	6.2×4.1×6.4	7.5×6.2×6.89	9.6×9.22×8.25	12.2×11.5×9.6	16.4×11.5×9.6	16×13.5×10	20.3×13.9×10	25×13.5×10	18×13×9.5	32×13×9.5

(3) Allair jig and airFlow jig

Allair jig and AirFlow jig are used for dry coal cleaning with dry jigging technology. Application of Allair jig or AirFlow jig has the advantage of no process water, which eliminates the need for fines dewatering and slurry confinement. Dry jigging eliminates the clean coal moisture penalty associated with wet processing. The first Allair® jig was commissioned in 2002 in the USA and more than 70 has been installed in recent years. Schematic diagram of Allair jig is shown in Figure 7.24.

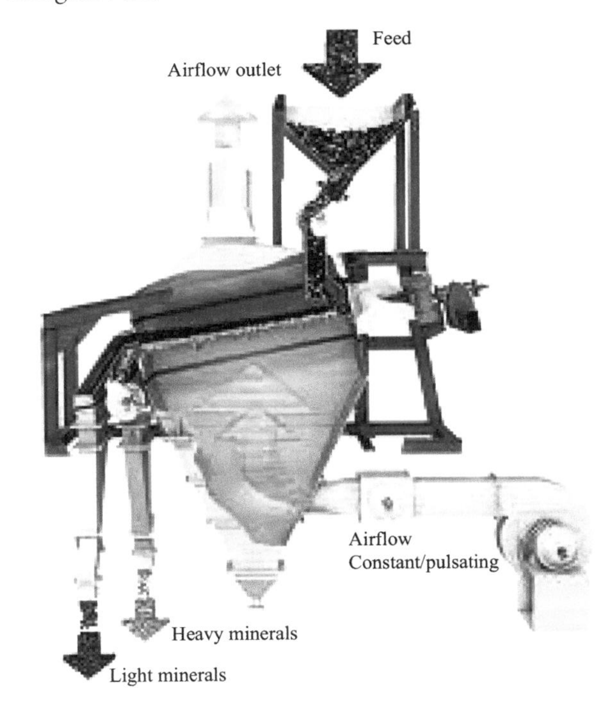

Figure 7.24 Schematic diagram of air magnetite separator

PROCESS EVALUATION AND PREDICTING COAL CLEANING RESULTS

The maximum recovery of cleaned coal at the required quality is of major concern to plant operators. However, the quantity and quality of products of coal cleaning predicted by theory based on washability studies and that obtained in the operation of actual commercial cleaning plants will be different for each coal. This chapter discusses the formulas used in coal separation evaluation and performance criteria in use today.

8.1 Determination of Clean Coal Yield by Product Ash Analysis

To determine the distribution curve for a particular coal cleaning operation, the following three tasks are required and/or must be performed; (1) clean coal yield, Yc, (2) float-and-sink analysis of the clean coal products, and (3) float-and-sink analysis of the refuse. The clean coal yield Yc can be determined either by direct weight, or it can be predicted by using material balance equations, if the average ash of feed coal, clean coal products, and refuse are known. A model for a two-product coal cleaning system is shown in Figure 8.1. A Mathematical representation of the mass balance in this cleaning system can be written as follows:

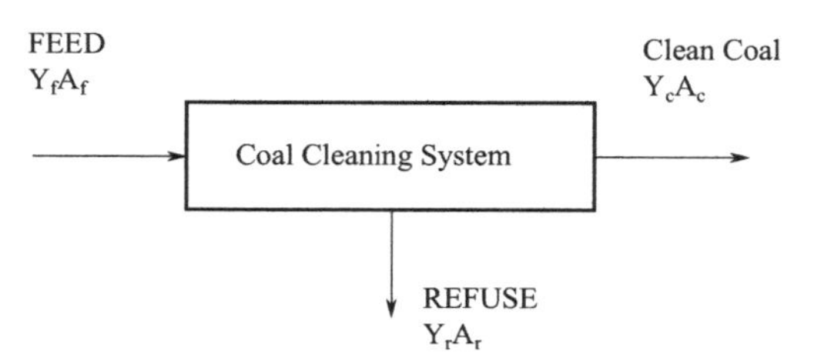

Figure 8.1 Simple model for a two-product coal cleaning system

$$A_cY_c+A_rY_r = A_fY_f \quad (8\text{-}1)$$

$$Y_c + Y_r = Y_f = 100 \quad (8\text{-}2)$$

where Y_f= feed coal = 100% wt% (feed basis);

Y_c= clean coal yield, wt%;

Y_r= refuse yield, wt%;

A_f= weighted average ash content of feed, wt%;

A_c= weighted average ash content of clean coal product, wt%;

A_r= weighted average ash content of refuse product, wt%.

This set of equations which include the ash balance Equation (8-1) and the yield balance Equation (8-2) can be solved simultaneously. The yield balance Equation (8-2) is multiplied by ($-A_r$), and add to the ash balance Equation (8-1). Rearrange the result, the yield of clean coal, Y_c is determined as follows:

$$Y_c = (A_r - A_f) / (A_r - A_c) \times 100\% \tag{8-3}$$

In practice, these equations are also used to adjust and control of the quantity and quality of clean coal products manually. By analysis of ash content of the coal samples obtained from feed, clean coal and refuse streams, the plant operator can change the specific gravity of separation.

8.2 Determination of Efficiency and Sharpness of Separation for a Coal Cleaning Unit

The following example is used to illustrate how to determine the efficiency and sharpness of separation of a coal cleaning unit.

A Baum jig having a rated capacity of 80 tons per hour cleans 2 by 0 inch bituminous coal. The jig has a total washing area of 90 square feet and washes coal at a rate of 100 tons per hour. The jig operates at 56 pulsations per minute and has a 12-inch stroke. Determine the following operational results of Baum jig operations:

① Throughput, tons per square foot of wash box area.

② Theoretical yield.

③ Actual yield.

④ Recovery efficiency.

⑤ Specific gravity of separation (SGS).

⑥ Probable error (EP).

⑦ Imperfection.

⑧ Amount of Float in the Refuse Product.

⑨ Amount of Sink in the Clean Coal Product.

⑩ The total amount of misplaced material.

Samples of the feed to the coal cleaning unit and clean coal and refuse products were taken over two operating shifts. These samples were float-and-sink tested and the specific gravity components analyzed separately for their ash contents. The results of the specific gravity analysis are given in Tables 8.1~8.3.

Table 8.1 Float-and-sink analysis of clean coal products

Specific gravity interval	Direct		Cumulative float	
	wt, %	Ash, %	wt, %	Ash, %
<1.30	53.2	6.5	53.2	6.5
1.30~1.40	40.3	12.6	93.5	9.1
1.40~1.60	5.5	26.4	99.0	10.1
1.60~1.80	0.5	44.0	99.5	10.3
>1.80	0.5	64.6	100.0	10.5

Table 8.2 Float-and-sink analysis of refused product

Specific gravity interval	Direct		Cumulative float	
	wt, %	Ash, %	wt, %	Ash, %
<1.30	2.0	6.1	2.0	6.1
1.30~1.40	7.1	4.4	9.1	12.6
1.40~1.60	16.9	30.2	26.0	24.0
1.60~1.80	16.5	48.3	42.5	33.5
>1.80	57.5	76.6	100.0	58.3

Table 8.3 Distribution of feed between clean coal product and refuse

Specific Gravity Interval	wt ① Percent			Percent Reporting to		Mean Specific Gravity
	Feed	Clean Coal	Refuse	Clean Coal	Refuse	
<1.30	47.90	47.70	0.20	99.60	0.40	1.28
1.30~1.40	36.90	36.20	0.70	98.10	1.90	1.35
1.40~1.60	6.60	4.90	1.70	74.30	25.70	1.50
1.60~1.80	2.15	0.45	1.70	19.00	81.00	1.70
>1.80	6.45	0.45	6.00	6.30	93.70	2.20
Total	100.0	89.7	10.3			

① recommended feed.

Solution:

① Throughput, tons per hour per square foot of separation area:

100 tons/hr/90 sq.ft. = 1.11 tons per hr per sq.ft. of separation area

② Theoretical recovery:

To determine the theoretical yield it will first be necessary to construct washability curves for the feed. The washability data is given in Table 8.4. The theoretical yield then is the yield of float read from the float yield-ash data in Table 8.4. For example, the cumulative float at 10.5 percent ash is approximately 91.6 percent by interpolating wt % vs. ash% in the cumulative float columns.

Table 8.4 Distribution of feed between clean coal products and refuse

Specific gravity interval	Direct ①		Cumulative float		Cumulative sink		Ordinate Z
	wt, %	Ash, %	wt, %	Ash, %	wt, %	Ash, %	
<1.30	46.5	6.6	46.5	6.6	100	15.4	23.2
1.30~1.40	38.1	12.4	84.6	9.2	53.5	23.1	65.5
1.40~1.60	7.2	27.1	91.8	10.6	15.4	49.7	88.2
1.60~1.80	1.8	47.4	93.6	11.5	8.2	69.5	92.7
>1.80	6.4	75.7	100	15.4	6.4	75.7	96.8
Total	100						

① Reconstituted feed.

③ Actual yield:

The actual yield is the percent of the feed actually recovered as the clean coal product. The recovery can be calculated on an ash basis by taking the material balance around the coal cleaning unit [see Equation (8-3)]. The actual yield is calculated from the following equation:

Actual Yield = $(A_r - A_f)/(A_r - A_c) \times 100\% = (58.3 - 15.4)/(58.3 - 10.5) \times 100\% = 89.75\%$

where A_f, A_c and A_r are the ash contents of the feed, clean coal, and refuse products. From Tables 8.4, 8.1 and 8.2, the ash contents of the feed, clean coal and refuse are 15.4, 10.5 and 58.3 percent, respectively.

④ Recovery efficiency:

The recovery efficiency by the Fraser-Yancey formula is:

Organic efficiency = (Actual yield of clean coal)/(Theoretical yield of clean coal at actual ash content) × 100%

Based on the engineering concept of efficiency, i.e., the ratio of the output to input, the recovery efficiency can be determined as

Recovery efficiency = (89.75/91.6) × 100 % = 98%

⑤ Specific gravity of separation (SG_{50} cut point, or d_{50}):

To determine the specific gravity of separation, (SGS), i.e., the specific gravity of the material in the feed which is equally distributed between clean coal and refuse products, a curve showing the distribution of the various specific gravity components of the feed between clean coal and refuse products must be constructed. By multiplying the weight of the clean coal and refuse in Tables 8.1 and 8.2 by their respective yields, and considering each specific gravity component as 100 percent, the distribution of each specific gravity component to clean coal and refuse can be calculated. These data are given in Table 8.3 and are plotted in Figure 8.2.

When plotting these data, each specific gravity component is plotted at its mean specific gravity. The meaning of a specific gravity fraction is the specific gravity of a bath in which half of the fraction would float and half would sink. These specific gravities may be determined from the arithmetic average of specific gravity intervals. The specific gravity at which the distribution curve crosses the ordinate representing equal distribution of the feed between clean coal and refuse products, namely, the 50 percent ordinate, is the specific gravity of separation (SG_{50}, cut-point, or d_{50}). Figure 8.2 shows a separation distribution curve of a coal cleaning device which gives 1.56 specific gravity across the 50 percent ordinate at SG_{50}.

⑥ Probable error (E_p):

On the assumption that the part of the separation distribution curve between the 25 and 75 percent ordinates in Figure 8.2 is a straight line, the probable error is the slope of this line. It may be determined by taking one-half the difference in the specific gravity at which the curve crosses the 25 and 75 percent ordinates.

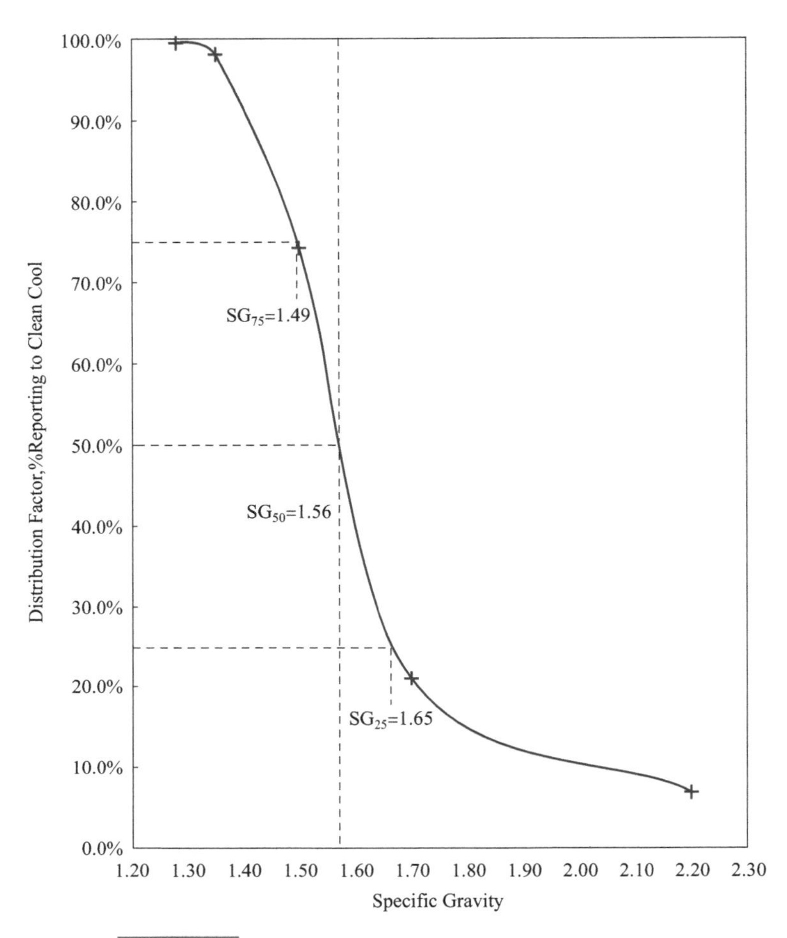

Figure 8.2 Distribution curve for coal cleaning unit

Probable error = $(SG_{25} - SG_{75})/2$

From Figure 8.2 the probable error is

Probable error = (1.65 − 1.49)/2 = 0.08

⑦ Imperfection (I):

The criterion for the sharpness of separation that is independent of density of separation is determined by.

I = Probable error / $(SG_{50} - 1)$ (Jig, Tabling, etc.)

I = Probable error / SG_{50} (Dense medium cleaning unit)

From Figure 8.2 for jig:

I = 0.08 / (1.56 − 1) = 0.143

⑧ Amount of float in the refused product:

The percent of float in the refuse may be interpolated from Table 8.3 by totaling all the material lighter than the specific gravity of separation (SG_{50}=1.56) shown to be in the refuse. In this particular case, it would be approximately:

0.2 + 0.7 + 1.7 = 2.6 (percent)

⑨ Amount of sink in the clean coal:

The percentage of sink in the clean coal product may also be interpolated from Table 8.3 by totaling all the material heavier than the specific gravity of separation shown to be in the clean coal product. In this case, it would be approximately the total amount of the clean coal weight% with SG heavier than 1.6:

0.45 + 0.45 = 0.9 (percent)

⑩ Total Amount of Misplaced Material:

The total amount of misplaced material is the sum of the float in the refuse, as calculated in step ⑧ and the sink in the clean coal, as calculated in step ⑨ , or

2.6 + 0.9 = 3.5 (percent)

8.3 Prediction of Coal Cleaning Results

The results of coal cleaning can be predicted from the results of the washability study by plotting the distribution curve (Figure 8.2) and presenting the data in a table as shown in Table 8.5, where computation among columns (a) to (g) are:

$e = b \times d/100$

$f = e \times c$

predicted clean coal yield% = summation of column e

predicted clean coal product ash% = summation column f

Table 8.5 Predicting coal cleaning results from distribution curve and washability data

Specific Gravity Interval	Direct (new raw coal)		Distribution factor reporting to product, %	Clean coal product		Mean specific gravity
	wt%	Ash%		wt%	Ash product, %	
(a)	(b)	(c)	(d)	(e)	(f)	(g)
<1.30	40	7	98.1	39.24	274.68	1.28
1.30~1.50	32	15	94.0	30.08	451.2	1.40
1.50~1.70	14	30	31.0	4.34	130.2	1.60
1.70~1.80	8	42	16.5	1.32	55.44	1.75
>1.80	6	50	6.3	0.38	19.0	2.20
	100			75.36	12.35	

Predicted Clean coal yield%=75.36%

Predicted clean coal ash%=12.35%

FROTH FLOTATION OF FINE COAL

Flotation, one of the important methods of coal processing, is a surface property based physical-chemical separation method. There are three phases involved: solid, liquid, and gas. In this chapter, the mechanisms of flotation, as well as the surface characteristics of coal will be discussed.

9.1 Surface and Interfaces of Solid-Liquid-Gas

① Quartz is a mineral composed of SiO_2. In water, the surface of the quartz forms the polar surface upon the breakage of primary chemical bond (covalent bond). The structure is shown below:

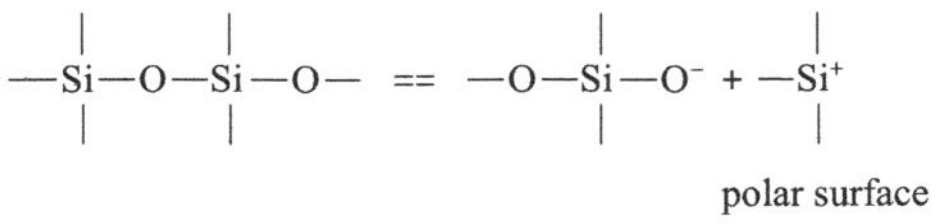

$$-Si-O-Si-O- == -O-Si-O^- + -Si^+$$

polar surface

This is dependent upon the pH of the solution:

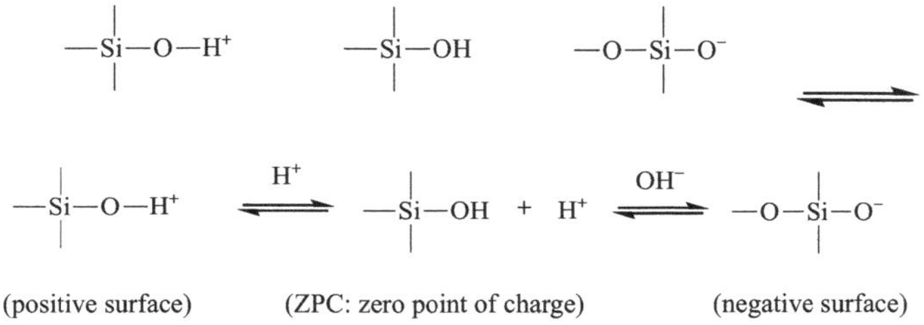

$$-Si-O-H^+ \qquad -Si-OH \qquad -O-Si-O^- \rightleftharpoons$$

$$-Si-O-H^+ \underset{}{\overset{H^+}{\rightleftharpoons}} -Si-OH + H^+ \overset{OH^-}{\rightleftharpoons} -O-Si-O^-$$

(positive surface) (ZPC: zero point of charge) (negative surface)

H^+ and OH^- are potential determined ions for the ash forming minerals such as SiO_2, Fe_2O_3, Al_2O_3, kaolinite, and insoluble oxides.

② Surface of coal upon breakage of the carbon - carbon surface forms the non-polar surface (the fracture of Van der Waals bonds).

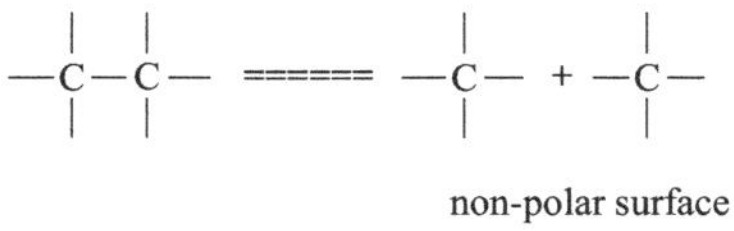

$$-C-C- ====== -C- + -C-$$

non-polar surface

The separation of fine coal is based on the surface property difference between coal (coal is nature's floatable material) and minerals. The surface of coal particles or coal-rich particles is attracted or strong affinity to oil or hydrophobic, while the surface of mineral particles or coal-lean particles has a strong affinity to water or hydrophilic. In other words, flotation process is to separate the solids based on the difference in wettability of particle surface.

In the froth flotation process of fine coal (Figure 9.1), coal, mineral matters, and air bubbles are mixed in a stirred tank flotation cell. If the density of air bubble with coal, $\delta_{(air + solid)}$, is less than that of water, δ_{water}, then the air bubbles with coal particles will rise to the surface of the water and are discharged as froth (clean fine coal), while mineral rich particles will settle in the bottom of the flotation cell and are discharged as tailings.

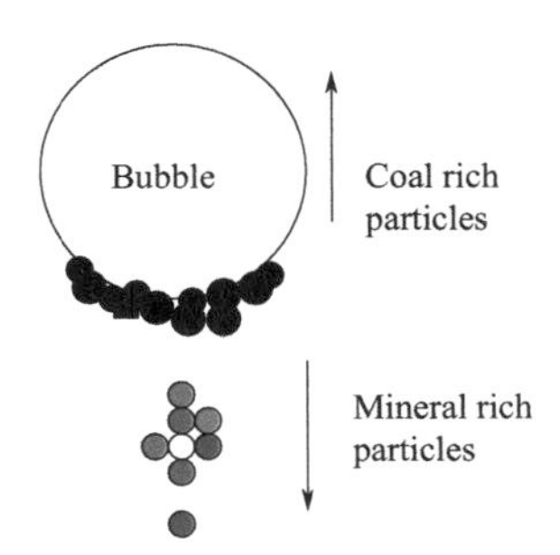

Figure 9.1 Coal rich and mineral rich particles in the froth flotation process

For flotation to occur, the free energy of the solid-liquid-gas (S-L-G) interfaces must decrease when a bubble is in contact with the solids at the minimal surface. At constant pressure (P) and temperature (T),

$$\gamma = \left(\frac{dG}{dA}\right)_{P,T} \tag{9-1a}$$

$$\Delta G = \gamma \Delta A \tag{9-1b}$$

where γ is surface tension, A is surface, and G is Gibbs free energy.

9.1.1 Young's Equation

Consider a gas bubble on a solid surface and submerged in a liquid forming a contact angle θ across the liquid. From the force vector as shown in Figure 9.2, the Young's equation can be derived as

$$\gamma_{GS} = \gamma_{LS} + \gamma_{LG} \times \cos\theta \tag{9-2}$$

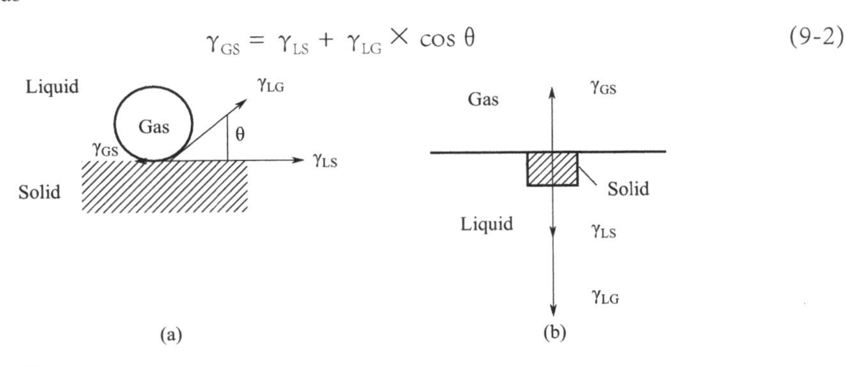

Figure 9.2 Force vectors in the Young's equation (a) and Dupree equation (b)

where γ_{GS} is the surface tension between solid and gas interface; γ_{LS} is the surface tension between liquid and solid interface; and γ_{LG} is the surface tension between liquid and gas interface.

9.1.2 Dupree's Equation

Consider a unit surface area of solid, the free energy per unit area corresponding to the attachment process is referred to as the Dupree's equation.

$$\Delta G = \left(\gamma_{GS} - \left(\gamma_{LG} + \gamma_{LS}\right)\right) \quad (9\text{-}3)$$

Substitute the Young's equation, Equation (9-2), into the Dupree equation to express the free energy change as a function of contact angle.

$$\Delta G = \gamma_{LS}\left(\cos\theta - 1\right) \quad (9\text{-}4)$$

where $\Delta G < 0$ for any finite value of contact angle, θ, and the process of solid attaching to gas is spontaneous for all finite contact angle.

9.2 Effect of Coal Rank and Mineral Inclusion on Coal Floatability

The rank of coal has different carbon content and shows different degrees of oxidation. Thus, the rank of coal shows different degree of contact angle and floatability (Figures 9.3 and 9.4, and Table 9.1). The mineral incursion in the coal affects the floatability of coal due to an increase in the density of the particles.

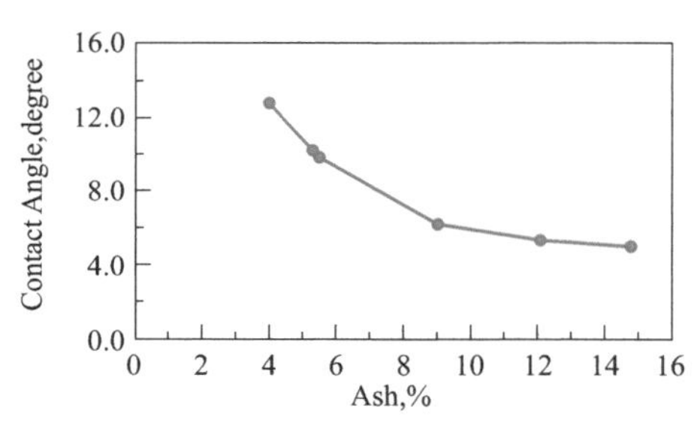

Figure 9.3 Contact angle as a function of ash content in coal

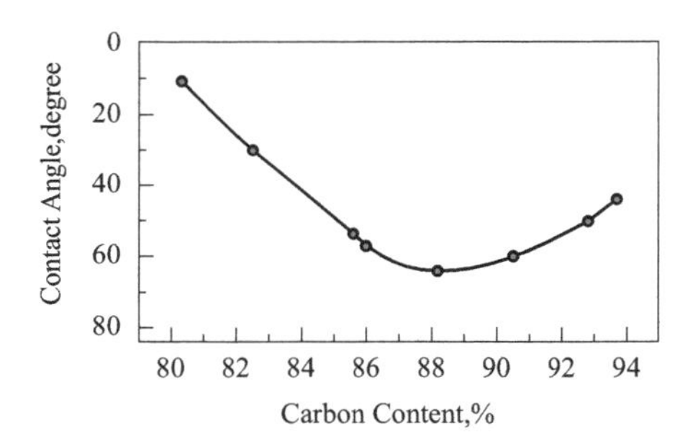

Figure 9.4 Contact angle as a function of the carbon content in coal

Table 9.1 Coal rank and floatability

Rank of coal	Ease of floatation
Bituminous (low volatile)	1
Bituminous (medium volatile)	2
Bituminous (high volatile)	3
Subbituminous	5
Anthracite	4
Lignite	6

9.3 Parameters Affecting the Rate of Flotation

As shown in Figure 9.5, flotation has three components, chemistry, equipment, and operation. The rate of flotation can be affected by the following factors:

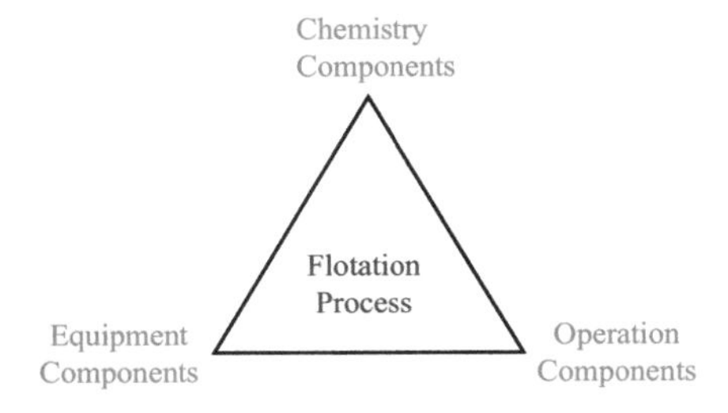

Figure 9.5 Three components in the flotation process

(1) Chemical components

① froth depth;
② frother concentration;
③ collector concentration;
④ pH.

(2) Operation components

① pulp density;
② particle size;
③ mineralogy;
④ feed rate;
⑤ wash water rate;
⑥ temperature.

(3) Equipment components

① cell design (see Figure 9.6);
② impeller speed (agitation);
③ aeration rate (air flowrate);
④ cell bank configuration and control.

The rate of flotation can be determined by:

① Probability of particle-bubble collision (p_c): The probability of collision between particles and bubbles in the flotation pulp phase;

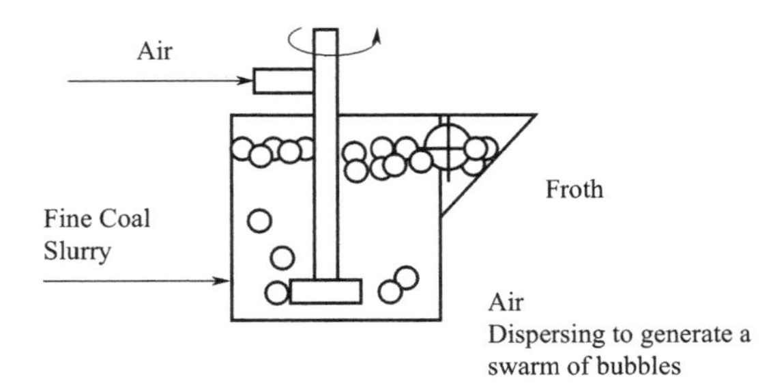

Figure 9.6 Stirred-tank type flotation cell

② Probability of particle-bubble adhesion (p_a): The attachment probability of particles and bubbles; Thinning and rupturing the water film at the air-solid interface; Induction time measurement: Bubble-particle attachment time determination.

③ Probability of particle-bubble detachment (p_d): The probability of bubbles breaks and drops back to the pulp in the pulp and drains back in the froth layer.

The flotation rate of particle size i can be calculated as

$$r_i = P_c\ P_a\left(1 - P_d\right)C_i V \tag{9-5}$$

where C_i = concentration of pulp and V = total volume of pulp

Specific flotation rate is

$$Q_i = r_i / \left(C_i \times V\right) \tag{9-6}$$

9.3.1 Rate of Flotation in Stirred Tank Cell Flotation

The rate of flotation is governed largely by the time available for attachment to occur (the nature of collision events) and the bubble attachment time between solids and gas phases in the aerated pulp. Accordingly, the rate of fine coal flotation is proportional to the hydrophobicity and concentration of coal particles as well as the surface properties of bubbles and availability of air bubble surface area per unit volume of pulp. Quantitatively, this can be described by the following first order rate equation for a given component.

$$-\frac{dW\left(\phi_{DP}, D_P, t\right)}{V} = N\ \phi_{DP} W\left(k, E_p, t\right) \tag{9-7}$$

where $W\left(k, E_p, t\right)$ is the mass of floatable particle of size, D_P and floatability, ϕ, that remains in the cell at time, t; N is the number of frother/collector coated air bubble per unit volume of pulp; V is total pulp volume.

For a given particle size, not all the coal particles are floatable under a specified set of operating conditions. Let define M as the mass fraction of the particles in the size D_P which are actually floatable.

By considering a mono-dispersed feed with a rectangular distribution function for distributed rate constant, and the first order rate equation for the flotation process, the amount of mass fraction of floatable particles that remain in the cell may be written for a specific floatable component, i, by the following expression:

$$T = M\left[1 - \int_{D_{P,\min}}^{D_{P,\max}} F(D_P,\phi) f_{D_P}(\phi)\, d\phi\, dD_P\right] \tag{9-8}$$

where

$F(D_P,\phi) = \exp(-\phi Nt)$ obtained from integrated first order rate law. N is the number of air bubbles and t is time of batch flotation. $f_{D_P}(\phi)$ is the distribution of flotability for each size fraction. This can be expressed as a standard distribution function (density function) such as a rectangular distribution, i.e., $f_{D_P}(\phi) = 1/\phi D_P$.

$$T_i = M_i \frac{1}{K_i t}\left[1 - \exp(-K_i t)\right] \tag{9-9}$$

$$R_i = 1 - T_i \tag{9-10}$$

Therefore,

$$R_{ci} = M_i\left[1 - \frac{1}{K_i t}\left(1 - \exp(-K_i t)\right)\right] \tag{9-11}$$

where K_i is the flotation rate constant, M_i is the ultimate recovery, and R_{ci} is the combustible material recovery. K_i and M_i for the kinetic model of the flotation of each component can be determined by the non-linear minimization computer program based on the conjugate gradient method or built in non-linear programming tools in PSIPLOT.

9.3.2 Pyrite Reverse Flotation Process for Desulfurization

Some flotation is floating the gangue instead of the valuable minerals. This process is called reverse flotation.

For coal desulfurization, the first stage is conventional flotation, while the second stage involves conditioning the first stage concentration with a dextrin type depressant. In other words, the coal is depressed and pyrite is recovered through the froth.

9.4 Weathered Coal (Oxidized Coal)

9.4.1 Sources of Weathered Coal

Coal can be oxidized when it is exposed to air, varying amounts of moisture, and bacterial action. The following coals can be considered to have varying degree of oxidation:

① Coal in pillar sections;

② Outcrop coal;

③ Surface-mined coal;

④ Stockpiled coal (after certain extended time);

⑤ Fine coal in tailing pond (after certain extended time).

The weathered coal shows a higher content of naturally occurring iron and sulfur reducing bacteria. In the presence of air and water, the iron and sulfur reducing bacteria (such as Thiobacillus ferrooxidans) begin to digest and convert the pyrite materials found so abundantly distributed in coal.

For pyrite sulfur, when it is oxidized, it releases the electrons in aqueous solution:

$$Fe^{2+} = Fe^{3+} + e^{-}$$

$$S^{2-} = SO_4^{2-} + 6\,e^{-}$$

Oxidation of pyrite (acid mine drainage formation) can occur in the following processes: (Thiobacillus ferrooxidans)

$$FeS_2 + 3\tfrac{1}{2}\,O_2 + H_2O = Fe^{2+} + 2\,SO_4^{2-} + 2\,H^{+} \quad (9\text{-}12)$$

$$Fe^{2+} + \tfrac{1}{4}\,O_2 + H^{+} = Fe^{3+} + \tfrac{1}{2}\,H_2O \quad (9\text{-}13)$$

$$Fe^{2+} + 3\,H_2O = Fe^{3+}(OH)_3 + 3\,H^{+} \quad (9\text{-}14)$$

$$FeS_2 + 14\,Fe^{3+} + 8\,H_2O = 15\,Fe^{2+} + 2\,SO_4^{2-} + 16\,H^{+} \quad (9\text{-}15)$$

9.4.2 Characteristics of Oxidized Coal

(1) Change in wettability

① reduction in water repellency (less hydrophobicity);

② decrease in contact angle.

(2) Change in floatability

The floatability of coal decreases due to

① increase in the non-floatable components, such as oxygen and ash;

② reduction in the floatable components, such as hydrogen and carbon.

(3) Chemical composition changes

The oxidized coal is completely lack of aromatic high boiling benzene-extractable oil that the unoxidized coal contains.

(4) Change in Zeta potential

Zeta potential is a measure of the strength of the charge present in the particles suspended in water. The oxidized coal is predominantly covered by the hydroxyl group (OH^-) and carboxylic group, which are acidic in character. These groups are negatively charged in water and their charges increase with increased degree of oxidation.

9.4.3 Adverse Properties of Oxidized Coal on Coal Preparation

① Oxidized coal is mechanically weakened and easily degraded due to an increase in micro-fractures. The surface of oxidized coal is severely cracked and resembles a dried mud cake.

② The specific gravity of oxidized coal is increased by the increase in oxygen and ash content of coal, while hydrogen and carbon on the oxidized coal surface are sharply reduced. This may cause the misplaced material to occur due to a poor gravity control. The heating value of coal also decreases.

③ The oxidized coal produces a large number of fines due to an increase in Hardgrove Grindability index, HGI.

④ The oxidized coal has a higher surface area due to an increase in fines. Therefore, a higher reagent dosage is required.

⑤ The surface hydrophobicity of oxidized coal decreases, thereby, rendering no flotation.

⑥ The decreases and release of cations into the solution, and the acids are formed on the coal surface. The acids are negatively charged in water and cause the pH to decrease.

The major soluble inorganic electrolyte ions released from coal are Fe^{2+}, Ca^{2+}, Mg^{2+}, Al^{3+}, etc.

9.4.4 Coal Preparation Methods of Controlling the Effects of Oxidized Coal

The adverse effects of oxidized coals can be eliminated by the following methods:

① Resurface the coal particles by chemical agents to make them more hydrophobic (Figure 9.7).

② Proper selection of additional flocculants to control the settling rate of coal fines and overflow clarity in thickener.

③ Make pH adjustment by feeding the slurry of hydrated lime, $Ca(OH)_2$, into the thickener clarified water pump. This will provide complete corrosion protection of the entire plant. It also adjusts the feed pH to near neutral with dissolved metallic ions to the oxidized state.

④ Additional equipment such as cyclone, screens, volume of tanks, froth cell, dewatering, etc. may be needed to bring existing process units back to their rated capacity.

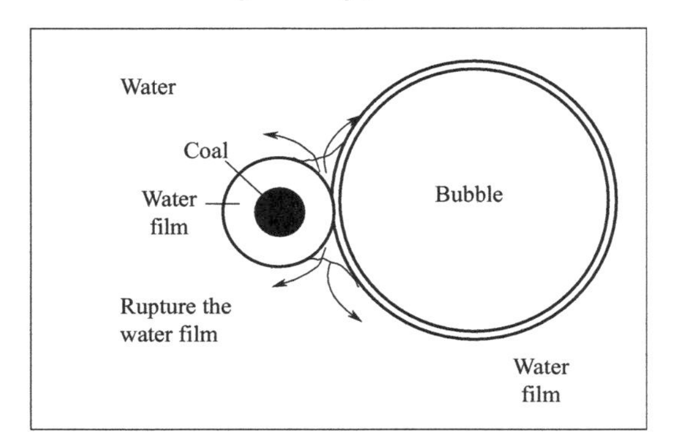

Figure 9.7 Break the water films to increase the affinity of coal particle to the air bubbles

9.5 Flotation Reagents

There are various types of reagents required for the flotation operation of coal. Table 9.2 shows the 6 types of reagents and their application in flotation of coal.

Table 9.2 Types of reagents used in coal flotation

① Frother - Foaming agents	MIBC, DOW M150, Pine oil, Cresylic acid.
② Collector - Surfactants or chemicals that adsorb on a solid to make it water repel	Kerosene, Fuel Oil No. 2 or No. 6
③ Modifier - Reagents that adjust the pH of solution	Hydrated lime [$Ca(OH)_2$], CaO, Soda ash Na_2CO_3, NaOH, HCl, H_2SO_4
④ Promoter - usually is blended with frothers to promote the adsorption of collectors	-
⑤ Depressant - Reagents that inhibit collector adsorption	Organic colloids such as Dextrin, Starch, Glue
⑥ Activator - Reagents that cause coal or minerals to adsorb a collector.	-

9.5.1 Effects of Non-Polar Hydrocarbon Oils

The non-polar hydrocarbon oils such as kerosene and fuel oil No.2 are used in flotation for the following purpose utilizing the fact that they enhance the adhesion of individual particle to a bubble:

① It increases the resulting contact angle of a coal against air bubbles substantially, over the value in its absence.

② It accelerates kinetics and makes possible the flotation of larger coal particle size.

9.5.2 Common Flotation Reagents Used in Fine Coal Flotation

(1) Frothers

Frother is a frothing reagent. It modifies the surface tension providing a stable and mobile froth in the flotation system. It is to reduce the surface tension of bubbles, preventing breaking of bubbles and bubble coalescence, and generating microbubbles. The common frothers for coal flotation are described below:

1) Aliphatic Alcohols

① R—OH R = C_4 to C_{10}

MIBC (methyl isobutyl carbinoal)

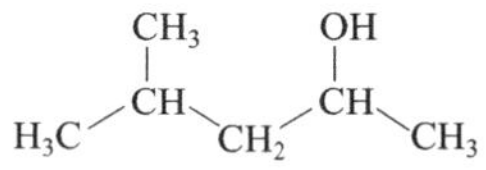

CH_3 OH
H_3C CH CH_2 CH CH_3

② 2-ethyl Hexanol

OH
H_3C CH_2 CH_2 CH_2 CH
CH_2
CH_3

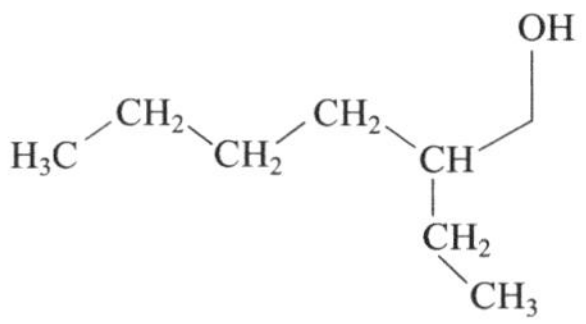

2) Ester alcohols

2,2,4 - Trimethylpentanediol (TEXANOL)

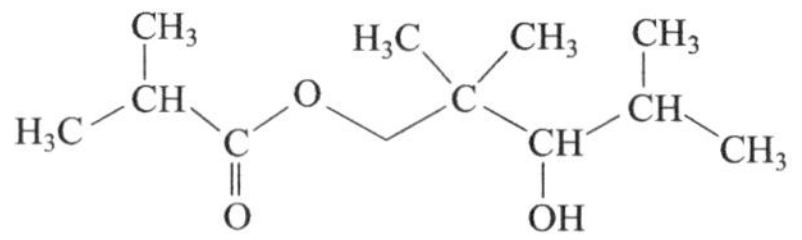

CH_3 H_3C CH_3 CH_3
H_3C CH C O C CH CH CH_3
O OH

Shur-Coal 168 (a blend of Texanol and a Sherex promoter (Keys 1986)

3) Polypropylene Glycol Methyl Ether

DOWFROTH Series: M252 (1263-D), M253(1012-D),

M254 (250-D), M255 (200-D)
(1012-D, n = 6.3)

$$H_3C \left(O - C_3H_6 \right)_n OH \quad n = 3 \text{ to } 7$$

4) Polypropylene Glycols
① Polypropylene Glycols
DOWFROTH series: M150, M222

$$H \left(O - C_3H_6 \right)_n OH \quad n = 3 \text{ to } 10$$

② AEROFROTH 65 for n=6.

$$H \left(O - \underset{CH_3}{CH} - C_2H_4 \right)_{6.5} OH$$

5) Pine Oil (α -Terpineol)

$$CH_3 - C_6H_{10} - C(CH_3)_2 - OH$$

6) Cresylic Acid (xylenol, o-Cresylic Acid)

$$2,6\text{-}(CH_3)_2C_6H_3OH$$

7) Alkoxy Paraffins
1,1,3 - Triethoxybutane (TEB)

$$CH_3CH_2-O-CH(OCH_2CH_3)-CH_2-CH(CH_3)-O-CH_2CH_3$$

Polyalkoxy Paraffins:

$$(OC_2H_5)_2CH-CH_2CH_2CH_2CH_2CH_3$$

(2) Collectors

Collectors are reagents that increase the hydrophobicity of the mineral surface, thereby improving the separability of the hydrophobic and hydrophilic particles. They increase the contact angle of a coal against air bubbles substantially and accelerate kinetics and make possible the flotation of larger coal particle size, both as a result of stronger adhesion of individual particle to the bubble. For coal flotation, kerosene and fuel oil are normally used.

① Kerosene is a combustible liquid which is derived from petroleum. The formula is $C_{12}H_{26}$—$C_{15}H_{32}$.

② Fuel Oil No.2 is obtained from the distillation of crude oil. It is approximately 75% aliphatic hydrocarbons ($C_{10}H_{20}$ - $C_{15}H_{28}$) and about 25% aromatic hydrocarbons (e.g. benzene, styrene).

(3) pH modifier

Flotation modifiers are used to adjust the conditions to increase the flotation performance. The common pH modifiers are:

① Lime (CaO).

② Soda ash (Na_2CO_3).

③ Caustic Soda (NaOH).

④ Sulfuric Acid (H_2SO_4).

⑤ Hydrochloric Acid (HCl).

9.6 Flotation Conditioner for Weathered Coal

DOWELL M210 is used specifically to improve recovery of hard to float coal fines. It is usually used in conjunction with DOWELL Frother M150 and diesel oil. Since it selectively conditions the coal, the ash content of recovered coal is maintained at a minimum level (Table 9.3).

Table 9.3 Effect of flotation of adding M210 with No.2 diesel fuel

M210/diesel fuel, lb/ton	M150 frother lb/ton	Slurry pH	Recovery, %	Product Ash, %
Pittsburgh seam - raw coal ash 14.4%				
4.0 diesel only	0.4	5.0	34.0	9.4
4.0 10 % M210 90 % diesel	0.4	5.0	87.0	9.7
4.0 10 % M210 90 % diesel	0.4	8.0	80.7	9.2
Printer seam - raw coal ash 23.5%				
0.8 diesel only	0.2	6.4	38.9	12.9
0.8 10 % M210 90 % diesel	0.2	6.3	94.6	13.4
Brookwood seam - raw coal ash 22.6%				

M210/diesel fuel, lb/ton	M150 frother lb/ton	Slurry pH	Recovery, %	Product Ash, %
0.37 diesel only	0.2	7.2	71.3	10.8
0.37 10 % M210 90 % diesel	0.2	7.2	89.6	10.9
Pittsburgh seam - raw coal ash 14.4%				
0.8 diesel only	0.2	9.0	77.3	5.9
0.8 10 % M210 90 % diesel	0.2	9.0	89.9	6.4

9.7 Calculation of Combustible Material Recovery

The combustible Material Recovery can be calculated by

$$R_c\% = 100\frac{Y_c(100 - A_c)}{(100 - A_f)}\% \tag{9-16}$$

where Y_c = Yield of clean coal product, %, A_c = Ash content of clean coal product, %, A_f = Ash content of feed, %, A_r = Ash content of refuse, %, and R_c = Combustible materials recovery %.

The combustible material recovery depends on flotation resident time and can be calculated by using Table 9.4 data

Combustible material recovery (CMR) % = $C(100 - A_c)/F(100 - A_f)$

$= Y_c(100 - A_{c)})/(100 - A_f)$ %

Table 9.4 Calculation of combustible material recovery

Flotation time, s	Individual		Cumulative			CMR
	Yield, wt%	Ash%	Yield, wt%	Product ash	Ash%	%
0						
15	11.71	10.22	11.71	119.6762	10.22	12.99
30	10.82	10.54	22.53	114.0428	10.37	24.95
60	18.5	10.36	41.13	192.696	10.37	45.56
120	24.1	10.78	65.23	259.798	10.52	72.13
180	8.78	11.39	74.01	100.0042	10.62	81.74
300	9.25	17.33	83.26	160.3025	11.37	91.19
Tailings	16.74	57.42	100.00	961.2108	19.08	100.00
				1907.7305		

Ash % of feed 19.08%

9.8 Unit Operations for Ultra-fine Coal

For ultrafine coal recovery, there are two types of flotation cells, stirred tank flotation cell and flotation column.

9.8.1 Stirred Tank Flotation Cell

The flotation cells consist of rotors and stators, which are also equipped with stainless steel components and air control valves for regulating air input. Below in Figure 9.8 is a depiction of a Denver D12 Laboratory Flotation Machine, a frequently utilized apparatus in laboratory flotation procedures.

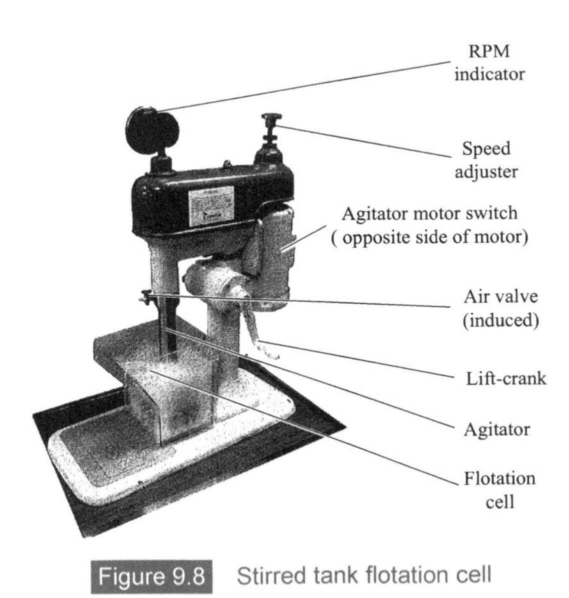

Figure 9.8 Stirred tank flotation cell

Figure 9.9 shows both the design and the operational mechanism of the flotation machine. There are three zones in the flotation cell: mixing and aeration zone, separation zone and concentrate zone.

Mixing Zone: The coal slurry is introduced into the cell by using gravity, and it descends directly onto the rotating impeller. Once the mixture cascades over the impeller blades, it experiences an outward and upward impetus, driven by the centrifugal force produced by the impeller's rotation. This mechanism utilizes the ejector principle, leading to a beneficial suction phenomenon. This suction efficiently pulls considerable and controlled volumes of air down the standpipe and into the central region of the cell.

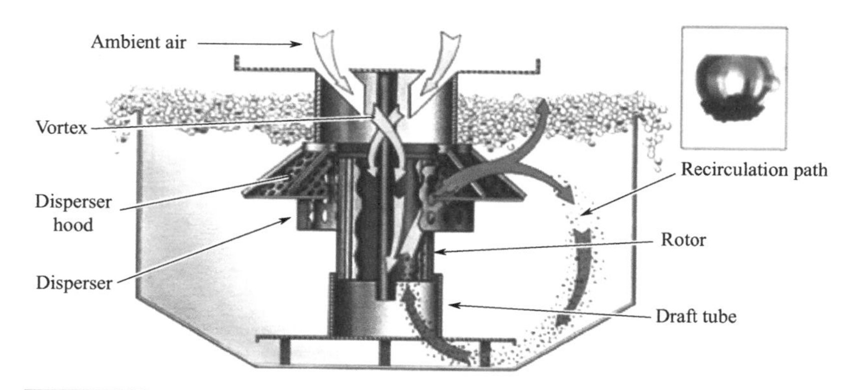

Figure 9.9 Unit operation for fine coal flotation (FLSmidth Krebs Manifold)

Separation Zone: Within the central region, a calm environment prevails, eradicating any intersecting currents. This effectively prevents coal particles from disengaging or colliding with the buoyant air bubbles that provide crucial support. In this area, the air-bubble-bound coal separates from the undesired gangue, and the intermediate product is directed back into the mixing zone.

Concentrate Zone: In the concentrate or upper zone, the coal undergoing enrichment encounters a degree of separation facilitated by a baffle, effectively isolating it from the concentrated discharge section of the apparatus. The environment within the cell at this point is notably serene, facilitating the smooth advancement of the coal concentrate and its swift extraction.

The image in Figure 9.10 illustrates the layout of the flotation plant.

Figure 9.10 Flotation plant

9.8.2 Flotation Column

Flotation columns offer a range of significant advantages that set them apart as a preferred choice in mineral processing. These advantages include the simplicity of their construction, achieved by eliminating the need for rotors and stator components. Through the implementation of counter-current flow, flotation columns enable the efficient separation of hydrophilic minerals. Their design also leads to low energy consumption, translating to cost savings over time, while their operational and maintenance costs remain notably low. The columns' distinctive feature of a greater height-to-diameter ratio enhances their performance, resulting in a deeper froth and increased chances for effective bubble-particle collisions. This extended retention time within the column augments the efficacy of particle recollection. Moreover, the heightened selectivity of flotation columns contributes to higher recovery rates, making them an attractive and efficient choice for mineral separation processes.

Figure 9.11 displays the design of a flotation column along with the corresponding illustrative drawing. Flotation columns are typically divided into two distinct zones: the collection zone and the froth zone.

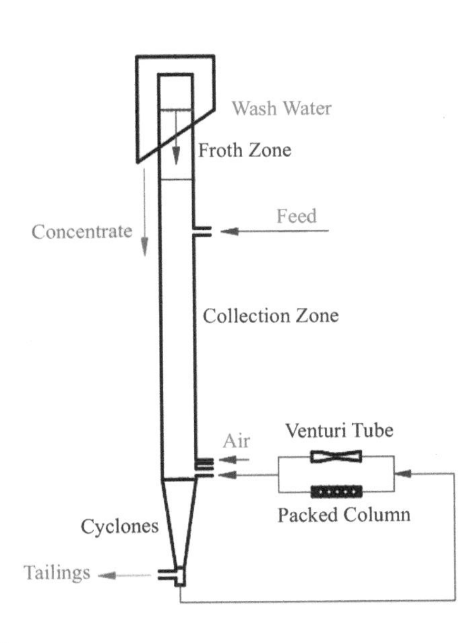

Figure 9.11 Lab flotation column

Collection Zone: In the absence of mechanical agitation, column cells lack a mixing zone. Instead, coal particles and bubbles maintain contact across the entire collection zone, extending from the spargers to the interface between the froth and the pulp. The coal slurry feed is introduced into the column near the upper part of the collection zone and descends through the ascending column of bubbles. Within the collection zone, bubble-particle aggregates are formed and subsequently transported upward to the froth zone.

Froth Zone: the froth zone in column cells boasts a significantly greater depth than that found in mechanical cells. This is attributed to the shape of the columns and is akin to the effect achieved by utilizing froth crowders in mechanical cells. The presence of washing water also contributes to stabilizing the froth by replenishing the water lost through gravity drainage. This dynamic enables a substantial degree of purification, to the extent that entrainment can be virtually eliminated.

Figure 9.12 illustrates the arrangement of flotation columns utilized in the industry.

Figure 9.12 Column floatation

DEWATERING, THICKENING, AND DRYING

10.1 Introduction

Dewatering is the removal of water from coal by sedimentation, filtration or similar solid-liquid separation processes. A thickening process is to increase the viscosity of a liquid or slurry without substantially changing other properties. Drying is a mass transfer process consisting of the removal of water from coal by evaporation.

Figure 10.1 shows the sedimentation and filtration techniques in mechanical dewatering. Equipment currently being used for mechanical dewatering of coal of varying sizes and final moisture content is illustrated in Figure 10.2.

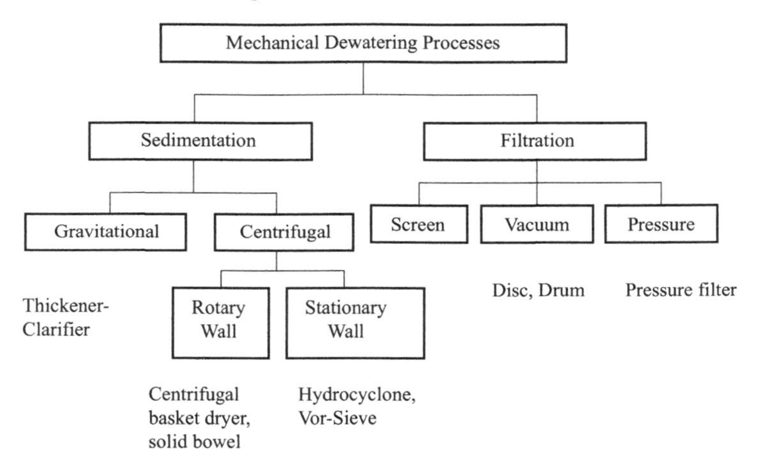

Figure 10.1 Sedimentation and filtration techniques in mechanical dewatering (Leonard, 1991)

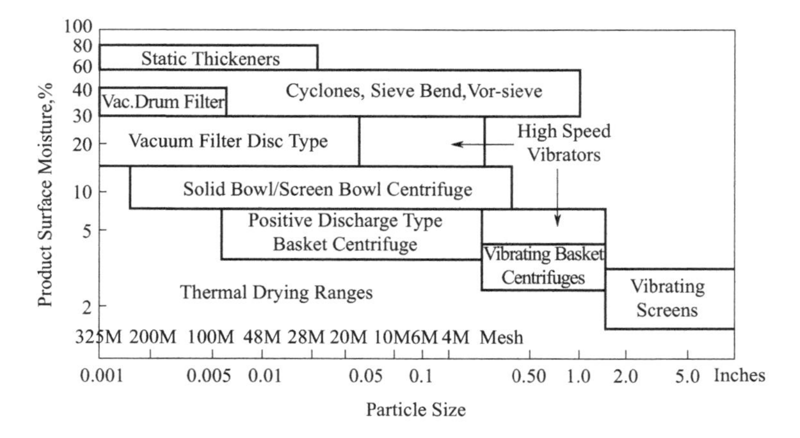

Figure 10.2 Equipment for dewatering of various particle sizes and approximate moisture in the product

10.2 Principle of Filtration (Cake formation in filtration)

In a filtration system as illustrated in Figure 10.3, the overall pressure drop distributions are

$$-\Delta p = p_a - p_b = (p_a - p') + (p' - p_b) = (-\Delta p_c) + (-\Delta p_m) \tag{10-1}$$

where $-\Delta p$ = overall pressure drop, $-\Delta p_c = (p_a - p')$ = pressure drop between inlet of cake and outlet of cake or pressure drop over the filter cake, $-\Delta p_m$ = pressure drop over filter medium, p_a = slurry inlet pressure, p' = cake outlet pressure, and p_b = medium outlet pressure.

In the low Reynolds number region, it is the viscous control, as the inertial force is unimportant. Thus, the friction or pressure drop flows through a bed of solids (in this case, the filtration cake) (Figure 10.3) can be expressed as follows from the Darcy's law and the Kozeny-Carman equation.

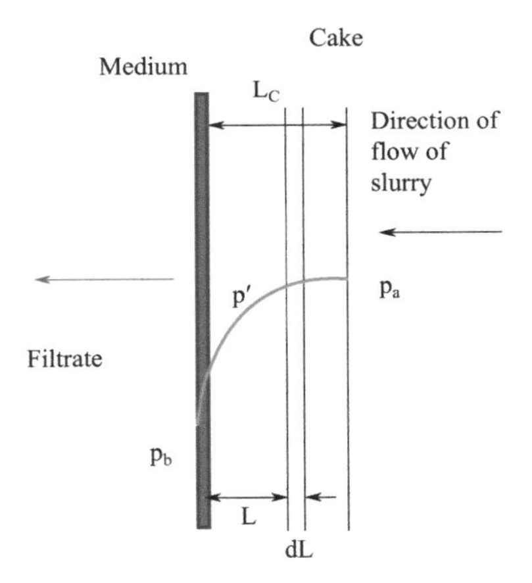

Figure 10.3 Diagram of pressure in a filtration system

Consider the pressure drop distribution for filtration:

$$\frac{-(p_a - p')}{\Delta L_c} = \frac{-dp_c}{dL} = k_1 \frac{(1-\varepsilon)^2 (\frac{S_p}{V_p})^2 \mu}{g_c \varepsilon^3} u + k_2 u^2 \tag{10-2}$$

(viscous force) (inertial force)

where

$-dp_c / \Delta L$ = pressure gradient in filtration cake of thickness L;

g_c = Newton's Law proportional fraction;

k_1= proportional constant for randomly packed particles $\approx$ 4.167;

k_2 = coefficient of inertial force;

L_c = thickness of flitter cake;

u = superficial or linear velocity of filtrate based on filtration area = $(dv/d\theta)/A$ where v is volume of filtrate, θ is dimensionless time, A is filtration area;

S_p = surface area of a single particle;

V_p = volume of a single particle;

ε = porosity of filtration cake (= 0.35 – 0.37) = volume of void /total volume;

μ = viscosity of filtrate.

10.2.1 Filter Cake Resistance, R_c

The volume of solid in the layer is $A(1-\varepsilon)\,dL$, and the density of particles is ρ_p, then the mass, dm of solids in the layer is

$$dm = A(1-\varepsilon)\rho_p\,dL \tag{10-2a}$$

Ignoring the inertial force in Eq (10-2) and integrating Eq (10-2), then plugging in Eq (10-2a), it becomes

$$-\int_{p}^{p'} dp = k_1 \frac{(1-\varepsilon)^2\left(\frac{S_p}{V_p}\right)^2}{\rho_p \varepsilon^3 A}\,\mu u \int_0^{m_c} dm \tag{10-2b}$$

$$\text{Let } \alpha = k_1 \frac{(1-\varepsilon)^2\left(\frac{S_p}{V_p}\right)^2}{\rho_p \varepsilon^3} \tag{10-3}$$

$$-\Delta p_c = \mu u\left(\frac{m_c}{A}\alpha\right) = \mu u R_c \tag{10-4}$$

Then

$$u = \frac{\frac{dV}{d\theta}}{A} \tag{10-5}$$

where

α = specific cake resistance;

$$\alpha = \alpha'(-\Delta p)^s \tag{10-6}$$

s = the compressibility coefficient of the cake;

s = 0 for incompressible slurry;

s = 0.2 ~ 0.8 for compressible slurry;

u = superficial or linear velocity of filtrate, based on filter area, A;

V = the volume of filtrate collected from the start of filtration to time θ;

R_c = filter cake resistance.

Cake resistance, R_c and filter medium resistance, R_m:

The specific cake resistance is defined as α as given in Eq (10-6).

$$-\Delta p = -(p_a - p_b) = \frac{\mu u}{g_c}(R_c + R_m) = \frac{\mu u}{g_c}\left(\frac{m_c}{A}\alpha + R_m\right) \tag{10-7}$$

where

m_c = Mass of solids in the cake;

R_m = Resistance of filter medium.

Let w_c = the mass of particles deposit in the filter per unit volume of filtrate.

Then, the mass of solid in the filter at time θ is

$$m_c = (\text{volume of filtrate})\left(\frac{\text{mass of filter cake}}{\text{unit volume of filtrate}}\right) = V w_c \tag{10-8}$$

Substitute Equations (10-4) and (10-5) into Equation (10-6) to give

$$\frac{d\theta}{dV} = \frac{\mu w_c \alpha}{A^2(-\Delta p) g_c}(V + A V_m) \tag{10-9}$$

where

$$V_m = \frac{\text{Fictitious volume of filtrate}}{\text{unit of filtering area necessary to lay down a fictitious cake thickeness}}$$

10.2.2 Filtration Equations

① For non-compressible slurry:

$$\frac{d\theta}{dV} = \frac{\mu w_c \alpha}{A^2(-\Delta p) g_c}(V + A V_m) \tag{10-9a}$$

② For compressible slurry:

$$\frac{d\theta}{dV} = \frac{\mu w_c \alpha'}{A^2(-\Delta p)^{1-s} g_c}(V + AV_m) \tag{10-9b}$$

where

A = filtration area;

g_c = Newton's Law proportional fraction;

Δp_c = pressure drop between inlet of cake and outlet of cake;

V = volume of filtrate collected from start of filtration to time, θ ;

w_c = the mass of particles deposit in the filter per unit volume of filtrate;

θ = filtration time;

μ = viscosity of filtrate;

α = specific cake resistance (i.e., it equals to the pressure drop required to give unit velocity of filtrate flow when the viscosity is unity and the cake contains one unit mass of solid per unit filter area) = $\alpha'(-\Delta p)^s$.

$$V_m = \frac{\text{Fictitious volume of filtrate}}{\text{Unit of filtering area necessary to lay down a fictitious cake thickeness}}$$

10.2.3 Methods of Filtration Operations

There are two methods of filtration, Constant Pressure Filtration and Constant Rate Filtrations:

(1) Constant pressure filtration

If Δp is constant, from Eq(10-9b)

$$\int_0^\theta d\theta = \frac{w_c(\alpha' \mu)}{A^2(-\Delta p)^{1-s} g_c}\int_0^V V\,dV + \frac{w_c(\mu\alpha')}{A(-\Delta p)^{1-s} g_c} V_m \int_0^V dV$$

$$\theta_{\Delta p} = \frac{w_c(\mu\alpha')}{2A^2(-\Delta p)^{1-s} g_c} V^2 + \frac{w_c(\mu\alpha') V_m}{A(-\Delta p)^{1-s} g_c} V \tag{10-10}$$

$$\theta_{\Delta p} = K_{1p} V^2 + K_{2p} V \tag{10-11}$$

where

A = filtration area;

θ = filtration time;

V = volume of filtrate collected from start of filtration to time, θ.

(2) Constant rate filtration

$$\frac{d\theta}{dV} = \frac{\mu w_c \alpha'}{A^2(-\Delta p)^{1-s} g_c}(V + AV_m) \tag{10-9b}$$

Re-arrange Eq (10-9b): $\frac{(-\Delta p)^{1-s}}{\alpha'} = \frac{\mu w_c}{A g_c}\left(V\frac{1}{A}\frac{dV}{d\theta} + AV_m \frac{1}{A}\frac{dV}{d\theta}\right)$

Substitute $\frac{dV}{Ad\theta} = \frac{1}{A}\frac{V}{\theta}$

to give

$$(\theta)_r = \frac{\alpha' \mu w_c}{2A^2(-\Delta p)^{1-s} g_c} V^2 + \frac{\alpha' \mu w_c V_m}{A(-\Delta p)^{1-s} g_c} V \tag{10-12}$$

$$(\theta)_r = K_{1r} V^2 + K_{2r} V \tag{10-13}$$

10.3 Filtration Experiment for Determination of Filtration Area

A filter leaf test apparatus is illustrated in Figure 10.4. The "leaf test" represents a small section of a continuous filter incorporating typical filter medium support and filtrate drainage. When used with techniques properly simulating the continuous filter, it can give results adequate for sizing production filters and predicting performance or determining the need for pilot testing.

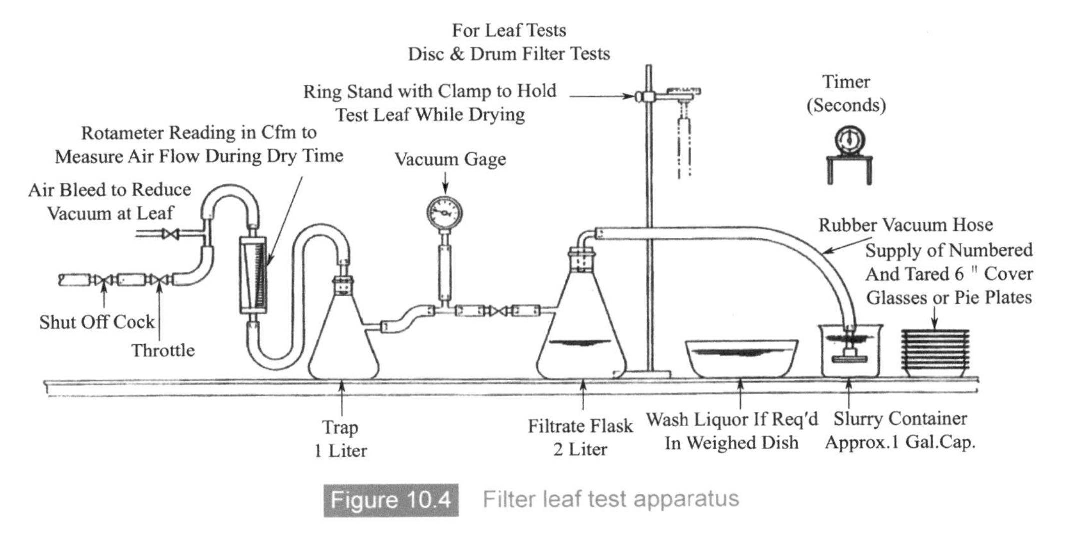

Figure 10.4 Filter leaf test apparatus

The leaf testing procedures are (911 Metallurgy Corp.):

① Stir the slurry with a spoon or spatula or an agitator to obtain a uniform suspension of the solids.

② Turn on the vacuum source and adjust the test vacuum while pinching off the hose to the test leaf.

③ Submerge the leaf in the slurry, open the hose and simultaneously start the timer. Stir the slurry as needed during the time.

④ At the end of planned form-time rotate the leaf up and out of the slurry just as a filter section and dry. During the dry time slowly rotate the leaf, tilting it from its horizontal plane to half drain the leaf. Dry for the planned time. Note the time of any cake cracking.

⑤ Turn off the vacuum, quickly remove the tubing from the flask to break the vacuum on the leaf. Elevate the leaf and drain the remaining filtrate to the flask. If necessary to facilitate draining, lift a portion of the cake off the cloth.

⑥ Explore the best method of cake discharge. If the cake is mushroom-shaped, extending beyond the 0.1 ft^2 cloth area, trim off and separately weigh the trim. Note: this correction is approximate.

⑦ Transfer the 0.1 ft^2 cake to a tared dish. Again allow the leaf to drain the residual filtrate to the flask. Set the leaf aside.

⑧ Measure and record cake thickness and any variations.

⑨ Weigh the dish and contents. Record the gross wet weight. Put the dish in an oven at 105℃ or lower temperature if necessary. Dry overnight, then record the gross dry weight.

10.3.1 Constant Pressure Filtration Experiment

K_{1p} and K_{2p} in Equation (10.11) can be determined from filtration experiment tests by plotting filtration time, θ/V vs. V. V is the volume of filtrate, as illustrated in Figure 10.5.

Knowing K_{1p} and K_{2p}, the following quantities for filtration system can be determined:

A = Filtration area,

α = Specific filter resistant, and

s = constant.

Example

Given: w_c = slurry solid concentration;

$-\Delta p$ = filtration pressure;

V = volume of filtrate;

θ = filtration time.

Solution: from Equation (10-10) and Equation (10-11)

$$\theta_{\Delta p} = \frac{w_c(\mu\alpha')}{2A^2(-\Delta p)^{1-s}g_c}V^2 + \frac{w_c(\mu\alpha')V_m}{A(-\Delta p)^{1-s}g_c}V \text{ and } \theta_{\Delta p} = K_{1p}V^2 + K_{2p}V,$$

then

$$\frac{\theta}{V} = K_{1p}V + K_{2p}$$

$$\frac{\theta}{V} = mV + b$$

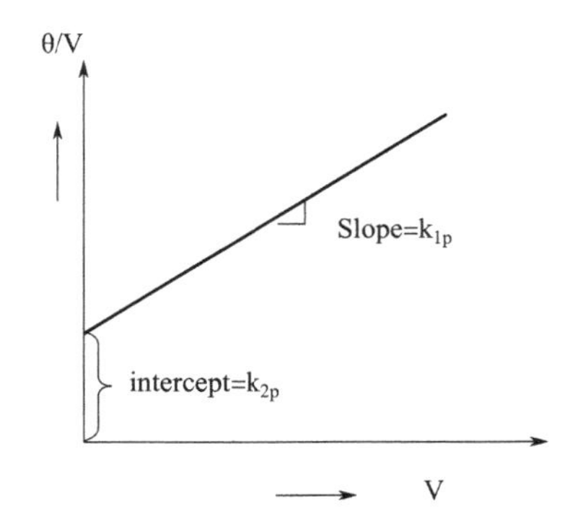

Figure 10.5 The ratio of filtration time to volume of filtrate vs. volume of filtrate

From K_{1p} and K_{2p} values other quantities of filtration system can be determined. The other parameters are: filtration area, A, thickness of cake, specific filter resistance, $\alpha = \alpha'(-\Delta p)^s$.

10.3.2 Estimation of Filtering Area Required for a Plate and Frame Filter Operation

A filter press is a tool used in separation processes, specifically in solid/liquid separation using the principle of pressure drive, provided by a slurry pump. The filter press is used in fixed-volume and batch operations, which means that the operation must be stopped to discharge the filter cake before the next batch can be started. As shown in Figure 10.6, the major components of a filter press are the skeleton and the filter pack. The skeleton holds the filter pack together while pressure is being developed inside the filtration chamber.

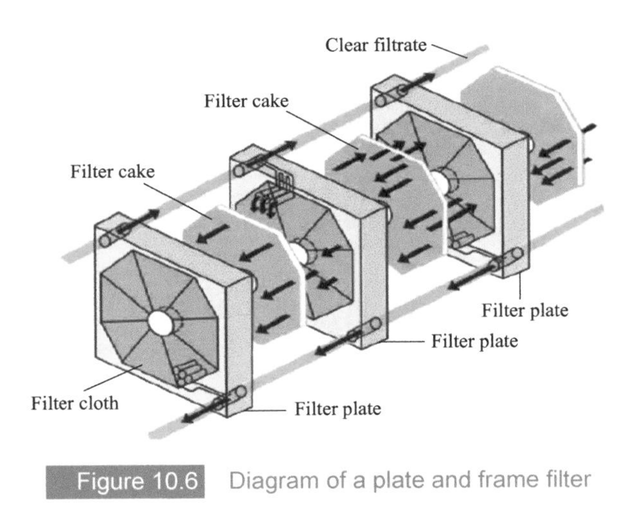

Figure 10.6 Diagram of a plate and frame filter

Example

Given: Slurry Concentration = $W_s = \dfrac{\text{lb of dry solid}}{\text{ft}^3 \text{ of liquid in the slurry}}$

μ of the liquid = 1 centipoise

If the filter must deliver at least 400 ft^3 solid free filtrate over a continuous operating time of 2 hours when the pressure difference driving force over the filter unit is constant at 25 psi. Based on the following data, obtained in a small plate-and-frame filter press and estimate the total area of filtering surface required.

The following experimental data was obtained from a plate-and-frame filter press with a total filtering area of 8 ft^3 (Table 10.1).

Table 10.1 Experimental data

Total volume of filtrated	Time from start of filtration, θ (hr), at constant pressure difference of		
V ft^3	$(-\Delta p) = 20$ psi	$(-\Delta p) = 30$ psi	$(-\Delta p) = 40$ psi
5	0.34	0.25	0.21
8	0.85	0.64	0.52
10	1.32	1.00	0.81
12	1.90	1.43	1.17

Solution:

In order to evaluate constants, V_m, S, and α', the following equation will be used:

$$(\theta)_{\Delta p} = \underbrace{\frac{\alpha' w_c \mu}{2A^2(-\Delta p)^{1-s} g_c}}_{K_{1p}} V^2 + \underbrace{\frac{\alpha' w_c \mu V_m}{A(-\Delta p)^{1-s} g_c}}_{K_{2p}} V$$

Rearrange the equation:

$$\frac{\theta(-\Delta p)}{V/A} = \frac{\alpha' w_c \mu (-\Delta p)^s}{2g_c}\left(\frac{V}{A}\right) + \frac{\alpha' w_c \mu (-\Delta p)^s}{g_c} \cdot V_m$$

Assuming the filtrate obtained was free of solid, and a negligible amount of liquid was retained in the cake, then

① For a given constant $-\Delta p$, a plot of $\frac{\theta(-\Delta p)}{V/A}$ vs. V/A should give a straight line with a

$$\text{Slope} = \frac{\alpha' w_c \mu (-\Delta p)^s}{2g_c}, \text{ and}$$

$$\text{Intercept} = \frac{\alpha' w_c \mu V_m (-\Delta p)^s}{g_c} \text{ (Figure 10.7)}$$

$-\Delta p$ $\left[\frac{lbf}{ft^3}\right]$	Slope $= \frac{\alpha' w_c \mu (-\Delta p)^s}{2g_c}$ $\left[\frac{hr \cdot lb}{ft^3}\right]$	Intercept $= \frac{\alpha' w_c \mu V_m (-\Delta p)^s}{g_c}$ $\left[\frac{hr \cdot lb}{ft^3}\right]$
20×144	2380	70
30×144	2680	80
40×144	2920	90

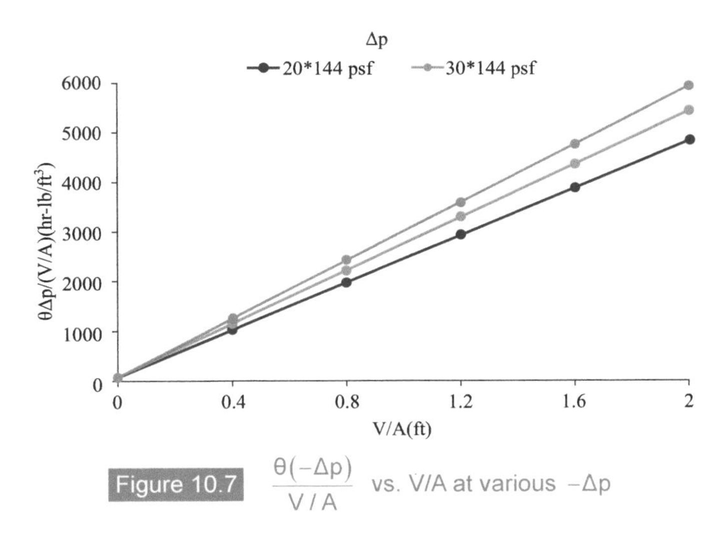

Figure 10.7 $\frac{\theta(-\Delta p)}{V/A}$ vs. V/A at various $-\Delta p$

② Take log of the slope and intercept,

$$\begin{cases} \text{Log (slope)} = s\log(\Delta p) + \log\left(\dfrac{\alpha' w_c \mu}{2g_c}\right) \\ \text{Log (intercept)} = s\log(\Delta p) + \log\dfrac{\alpha' w_c \mu V_F}{g_c} \end{cases}$$

A log − log plot of (slope) vs. Δp should give a straight-line with a slope of S and intercept of ($\frac{\alpha' w_c \mu}{2g_c}$).

From the slope, S= 0.3, and

From the intercept $\frac{\alpha' w_c \mu}{2g_c} = 220$,

Similarly, a log-log plot of (Intercept) vs. Δp should give a straight line with a slope of S and an intercept of ($\alpha' w_c \mu V_F / g_c$) (see Figure 10.8). Because the value of V_F is relatively small, the value of V_F will be estimated from the combined result of Figures 10.7 and 10.8.

Since W = slurry concentration = 5 lb/ft^3, and

μ = 2.42 lb/hr-ft

Therefore, $\alpha' = 220\frac{2g_c}{w_c\mu} = 220\frac{2g_c}{(5)(2.42)} = 36\left[\frac{hr^2}{lb}\right]$

From Figure 10.7, Intercept = 70 for $\Delta p = 30 \times 144\left[\frac{lb}{ft^2}\right]$

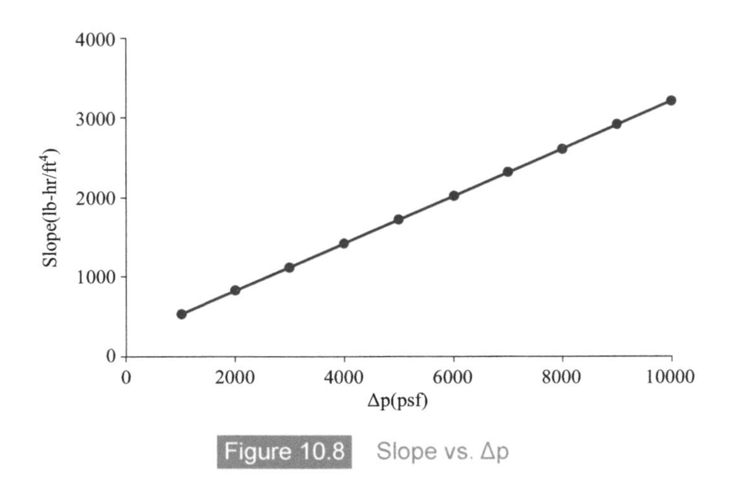

Figure 10.8 Slope vs. Δp

$$V = \frac{70}{\alpha' w_c \mu \ (\Delta p)^s} = \frac{70}{(36)(5)(2.42)(30 \times 144)^2} = 0.015\left[\frac{lb^3}{ft^2}\right]$$

Therefore, the final equation for use in evaluating the total filtrating area needed for the large filter is given by

$$(\theta)_{\Delta p} = \frac{36 \text{ w } \mu}{2A^2(\Delta p)^{1-0.3}} V^2 + \frac{36 \text{ w } \mu \ (0.015)}{A(\Delta p)^{1-0.3}} V,$$

$$(\theta)_{\Delta p} = \frac{18 \text{ w } \mu}{A^2(\Delta p)^{0.7}} V^2 + \frac{0.54 \text{ w } \mu}{A(\Delta p)^{0.7}} V$$

③ For the conditions of this problem case

$$V = 400 \ [ft^3]$$

$$\Delta p = 25 \times 144\left[\frac{lb}{ft^2}\right]$$

$$W = 5\left[\frac{lb}{ft^3}\right]$$

$$\mu = 2.42\left[\frac{lb}{ft - hr}\right]$$

$$\theta = 2[hrs]$$

Substitute the above values into the final equation,

$$2 = \frac{(18)(5)(2.42)}{A^2(25\times144)^{0.7}}(400)^2 + \frac{(0.54)(5)(2.42)}{A \bullet (25\times144)^{0.7}}(400)$$

Solving for A,

$$A = 240\ ft^2$$

Therefore, the total area of filtering surface required is approximately 240 sq. ft.

10.4 Continuous Filtration Operations

There are two types of filters used for constant filtration, rotary drum filter and a rotary disc filter.

A rotary vacuum filter drum as shown in Figures 10.9 and 10.10 which consists of a cylindrical filter membrane that is partly sub-merged in a slurry to be filtered. The inside of the drum is held lower than the ambient pressure. As the drum rotates through the slurry, the liquid is sucked through the membrane, leaving solids to cake on the membrane surface while the drum is submerged. A knife or blade is positioned to scrape the product from the surface.

Like the rotary vacuum drum filter, a disc vacuum filter as shown in Figure 10.11 pulls slurry from a tank under the center barrel of the filter. A solid cake forms on both sides of the disc surface while submerged. As the disc rotates out of the slurry tank, the sector continues pulling vacuum, allowing the cake to dry.

The dewatering process in continuous filtration operations consists of four steps in series:

① Cake formation,

② Washing (option),

③ Drying, and

④ Scraping.

The pressure drop across the filter during the cake formation is constant. From Eq (10-10)

$$\theta_{\Delta p} = \frac{w_c \mu \alpha'}{2A^2(-\Delta p)^{1-s} g_c} V^2 + \frac{w_c \mu \alpha' V_m}{A(-\Delta p)^{1-s} g_c} V \tag{10-10}$$

$$\theta_{\Delta p} = \frac{1}{N_R} = \frac{w_c \mu \alpha'}{2 A_D^2\ \phi_f\ (-\Delta p)^{1-s} g_c} V_R^2 + \frac{w_c \mu \alpha' V_m}{A_D(-\Delta p)^{1-s} g_c} V_R \tag{10-14}$$

let $K_{1c} = \dfrac{w_c \mu \alpha'}{2 A_D^2\ \phi_f(-\Delta p)^{1-s} g_c}$ and $K_{2c} = \dfrac{w_c \mu \alpha' V_m}{A_D(-\Delta p)^{1-s} g_c}$, then

$$\theta_{\Delta p} = \frac{1}{N_R} = K_{1c} V_R^2 + K_{2c} V_R \tag{10-15}$$

where A = total available area for continuous rotary drum filter;

ϕ_f = the fraction of total available area immersed in the slurry;

A_D = effective area of the filtering surface;

N_R = number of revolution per unit time.

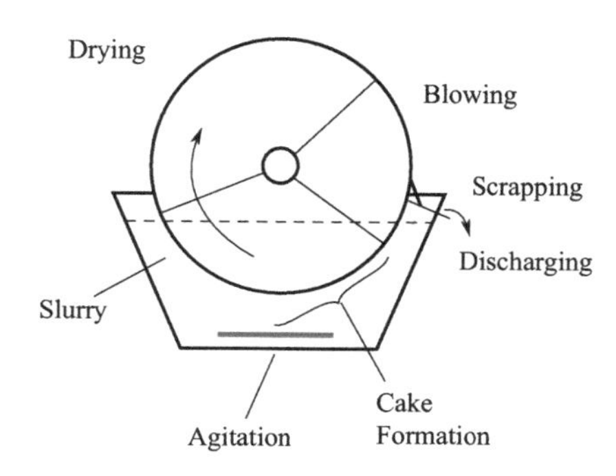

Figure 10.9 Diagram of continuous filtration process

In a continuous filtration system, as illustrated in Figure 10.9, assuming $R_c >> R_m$, i.e., R_m is negligible compared to R_c, and if the material is granular (rigid) solid, the filter cake can be assumed as incompressible. Then

$$\text{Volume of filtrate per revolution} = V_R = A_D\left[\frac{2\,\phi_f\,(-\Delta p)^{1-s}}{(\mu\,\alpha')\,w_c\,N_R}\right]^{0.5} \tag{10-16a}$$

$$\text{Volume of filtrate per unit time} = V_R\,N_R = A_D\left[\frac{2\,\phi_f\,(-\Delta p)^{1-s}}{(\mu\,\alpha')\,w_c}\right]^{0.5} \tag{10-16b}$$

$$\text{Weight of dry cake per unit time} = V_R\,N_R\,w_c = A_D\left[\frac{2\,\phi_f(-\Delta p)^{1-s}}{\mu\,\alpha'}\right]^{0.5} \tag{10-16c}$$

In a continuous rotary drum or disc filter, vacuum is the driving force that is employed to achieve the differential pressure across the cake. Air suction rate is defined by cubic feet of air per minute pulled through the cake during dewatering.

The flowing describes determination of air suction rate in rotary vacuum filter.

A vacuum pump must be supplied for the operation of a rotary vacuum filter. To estimate the size of pump and power requirement for a given filtering unit, the rate at which the air is sucked through the dewatering section of a rotary vacuum filter can be expressed as

$$\theta_a = \frac{1}{N_R} = \frac{w_c(\mu_a\beta')}{2A_D^2\,\phi_a\,(-\Delta p)^{1-r}\,g_c}V_a\,V_{aR} + \frac{w_c(\mu\beta)V_m}{A_D(-\Delta p)\,g_c}V_{aR} \tag{10-17}$$

$$\theta_a = K_{1a}V_R\,V_{ar} + K_{2a}V_{aR} \tag{10-18}$$

where A = total available area for continuous rotary drum filter;

ϕ_f = effective area of the filtering surface;

N_R = number of revolution per unit time;

V_a = volume of air at the temperature and pressure of surroundings suction in the time θ;

β = specific air suction cake resistance. If slurry is compressible, then

$\beta = \beta'(-\Delta p)^{r}$;

φ_a = the fraction of total available for air suction;

μ_a = viscosity of air at the temperature and pressure of surrounding.

In continuous filtration system, (1) $R_c >> R_m$, i.e. R_m is negligible as compared to R_c and (2) if the material is granular (rigid) solid, the filter cake can be assumed to be incompressible.

From Equations (10-16a) and (10-16b),

$$\text{Volume of air per revolution} = V_{aR} = A_D^2 \frac{2\phi_a(-\Delta p)}{(\mu_a\beta)\, w_c V_R N_R} \tag{10-19a}$$

$$\text{Volume of air per unit time} = V_{aR}\, N_R = A_D^2 \frac{2\phi_a(-\Delta p)}{(\mu_a\beta)\, w_c\, V_R} = \frac{A_D^2\phi_a}{\mu_a\ \beta}\left[\frac{2(-\Delta p)}{w_c V_R}\right] \tag{10-19b}$$

$$\frac{\text{Volume of air per unit of time}}{\text{Weight of dry cake per unit time}} = \frac{\phi_a}{\phi_f}\frac{\mu}{\mu_a}\frac{\alpha}{2\beta w} \tag{10-19c}$$

where A = total available area for continuous rotary drum filter;

ϕ_f = the fraction of total available area immersed in the slurry;

$A\phi_f$ = effective area of the filtering surface;

N_R = number of revolution per unit time.

Example:

A rotary drum filter with 30% submergence is used to filter a concentrated slurry of $CaCO_3$ containing 14.7 lb of solid per ft^3 of H_2O. The pressure drop is 20 in Hg. If the filter cake contains 50% moisture (wet basis), calculate the filter area required to filter 10 gal/min of slurry when the filter cycle time is 5 min.

Assuming that the specific cake resistance is given by

$$\alpha = 2.08\times10^{10}(+\Delta p)^{0.3}\left[\frac{ft}{lb}\right]\left[\frac{lbf}{ft^2}\right]$$

and the filter medium resistance R_F is negligible.

Solution:

First calculate the pressure difference and filtration time in one filter cycle:

$$\Delta p = (20\text{ in Hg})\frac{14.69\text{ psi}}{(29.92\text{ in Hg})} \times 144\frac{in^2}{ft^2} = 1{,}414\left[\frac{lb_f}{ft^2}\right]$$

$$\theta = \varphi_f \bullet \theta_t = (30\%)(5 \times 60) = 30\%(300\ \text{sec})$$

where θ = time required for formation of the cake;

θ_t = total cycle time = 5 minutes;

φ_f = fractional submergence of the drum in the slurry = 30%.

Correction of retention of liquid in the cake:

$$W = \frac{W_s}{1 - \left[\frac{M_F}{M_K} - 1\right]\frac{w_s}{\rho}}$$

where M_F = the mass of the wet cake including the filtrate retained in its voids;

M_K = mass of dry cake obtained by washing the cake of solvable material and drying;

ρ = density of filtrate;

W_s = concentration of solid in the slurry $\left[\frac{\text{lb}}{\text{ft}^3\ \text{of liquid fed to the filter}}\right]$,

$$W = \frac{\text{mass of particles deposited on the filter}}{\text{volume of filtrate}}$$

$$\frac{M_F}{M_K} = 2 \qquad (50\%\ \text{moisture in cake})$$

$$W_S = 14.9\frac{\text{lb}}{\text{ft}^3}; \qquad \rho = 62.3\frac{\text{lb}}{\text{ft}^3}$$

$$\mu = 1\text{cp} = 6.72 \times 10^{-4}\ \frac{\text{lb}}{\text{ft} - \text{sec}}$$

$$W = \frac{14.7}{1 - (2-1)\frac{14.7}{62.3}} = 19.24\left[\frac{\text{lb}}{\text{ft}^3}\right]$$

Use the equation: (assume $R_F \ll R_K$)

Material Balance:

$$\theta = \frac{1}{N_R} = \frac{\propto' \mu\, w}{2A_D^2 \varphi_f (\Delta p)^{1-s}} V_R^2 \qquad \frac{V_S}{\theta_t} \times w_S = \frac{V_R}{\theta_t} \bullet W$$

$$\text{Therefore, } A_D = \frac{V_R}{\theta_t}\left[\frac{\propto' w \cdot \mu \cdot \theta_t}{2\varphi_f (\Delta p)^{1-s}}\right]^{\frac{1}{2}} \qquad \frac{V_S}{\theta_t} \longrightarrow \Big| \longrightarrow \frac{V_R}{\theta_t}$$

$$\text{The filtrate rate} = (\text{feed rate})\left(\frac{W_s}{W}\right) = \left(\frac{V_S}{\theta_t}\right)\left(\frac{W_s}{W}\right)$$

$$\frac{V_R}{\theta_t}=\frac{10\ \text{gal / min}}{60\ \text{sec / min}}\times\frac{1}{7.48\frac{\text{gal}}{\text{ft}^3}}\times\frac{14.7\frac{\text{lb}}{\text{ft}^3}}{19.24\frac{\text{lb}}{\text{ft}^3}}=0.0170\left[\frac{\text{ft}^3}{\text{sec}}\right]$$

$$A_D=0.0170\ \frac{\text{ft}^3}{\text{sec}}\left[\frac{\left(2.08\times10^{10}\ \frac{\text{ft}}{\text{lb}}\right)\left(19.24\frac{\text{lb}}{\text{ft}^3\ \text{filtrate}}\right)\left(6.72\times10^{-4}\ \frac{\text{lb}}{\text{ft}-\text{sec}}\right)(300\ \text{sec})}{2(30\%)(1414\frac{\text{lbf}}{\text{ft}^3})^{1-0.3}\left(32.192\frac{\text{lb}-\text{ft}}{\text{lb}_f-\text{sec}}\right)}\right]^{\frac{1}{2}}$$

$$=86.9\ \text{ft}^2$$

10.5 Centrifugal Filtration

Centrifugal dryers can create high gravity forces in coal dewatering. In a centrifuge, centrifugal is the driving force to create a differential pressure in order to obtain fluid flow through the filtering medium. Centrifuges have wide application in wet washing coal plants because they are reliable and efficient machines and the filtration products are consistent, uniform, and easily handled (Leonard, 1991).

From Equation (10-9)

$$\frac{dV}{d\theta}=\frac{A^2(-\Delta p)^{1-s}g_c}{\alpha'\ \mu\ w_c\left(\frac{V}{A_a\ A_f}+\frac{V_m}{A_f}\right)} \tag{10-20}$$

where A_f = area of filter medium (inside area of centrifugal basket)=$2\ \pi\ r_2 b$,

A_a = the arithmetic mean cake area = $(r_i+r_2)\ \pi\ b$,

A_L = logarithmic mean cake area.

$$A_L=\left(2\ \pi\ b/(r_2-r_i)/\ln\left(\frac{r_2}{r_i}\right)\right) \tag{10-21}$$

b = height of the centrifugal basket,

r_2 = inside radius of the basket,

r_i = inner radius of the cake face,

r_1 = inner radius of the liquid surface.

The highlighted are the changes.

$$\text{Superficial or linear velocity of filtrate}=\frac{\frac{dV}{d\theta}}{A}=\frac{Q}{A} \tag{10-22}$$

Since the pressure drop $(-Cp)$ is due to centrifugal action,

$$-\Delta p = \frac{\rho\ \omega\left(r_2^2 - r_1^2\right)}{2\,g_c} \tag{10-23}$$

where Q = volumetric flowrate of filtrate,

ω = angular velocity (rad/sec),

ρ = density of the filtrate or liquid.

Then,

$$\frac{dV}{d\,\theta} = \frac{\left[\rho\ \omega\left(r_2^2 - r_1^2\right)\right]^{1-s} g_c}{2\,\alpha'\,\mu\,w_c g\left(\dfrac{V}{A_a\,A_f} + \dfrac{V_m}{A_f}\right)} \tag{10-24}$$

Integrating at constant pressure from $\theta = 0, V = 0$ to $\theta = \theta, V = V$ gives

$$\int_0^{\theta} d\,\theta = \frac{2\,\alpha'\ \mu\ w_c g}{\left[\rho\ \omega\left(r_2^2 - r_1^2\right)\right]^{1-s} g_c}\left(\frac{1}{A_a A_f}\int_0^{V} V\,dV + \frac{V_m}{A_f}\int_0^{V} dV\right)$$

$$\theta_f = \frac{4\,\alpha'\ \mu\ w_c g}{A_a A_f\left[\rho\ \omega\left(r_2^2 - r_1^2\right)\right]^{1-s} g_c} V^2 + \frac{2\,\alpha'\ \mu\ w_c g\, V_m}{A_f\left[\rho\ \omega\left(r_2^2 - r_1^2\right)\right]^{1-s} g_c} V \tag{10-25}$$

$$\theta_f = K_{1f} V^2 + K_{2f} V \tag{10-26}$$

10.6 Vacuum Filters

There are three types of continuous vacuum filters used in the coal industry, rotary drum, rotary disk, and horizontal belt or disk.

The rotary drum filter is the most widely used continuous vacuum filter. As shown in Figure 10.10, it consists of a cylindrical drum supported in an open tank. The drum shell has compartments covered with drainage grid or filter cloth. Each compartment automatically applied either suction or positive air pressure as the drum rotated through the slurry (Leonard, 1991).

Rotary-disk vacuum filters are also known as sector disc filters (Micronics Engineered Filtration Group Inc). The disc filters provide maximum filtration area in minimum space at minimum cost. In this filter, several discs stand in front of one another, connected to a center barrel and rotating through the slurry, open air, and the discharge zone. Each disc is comprised of sectors, shaped like a slice of pizza. Each sector is either grooved or made of punch plate to allow liquid to pass through a cloth bag, into the sector, and out the center barrel. Both sides of each disc filter slurry simultaneously. A rotary disc filter is shown in

Figure 10.11.

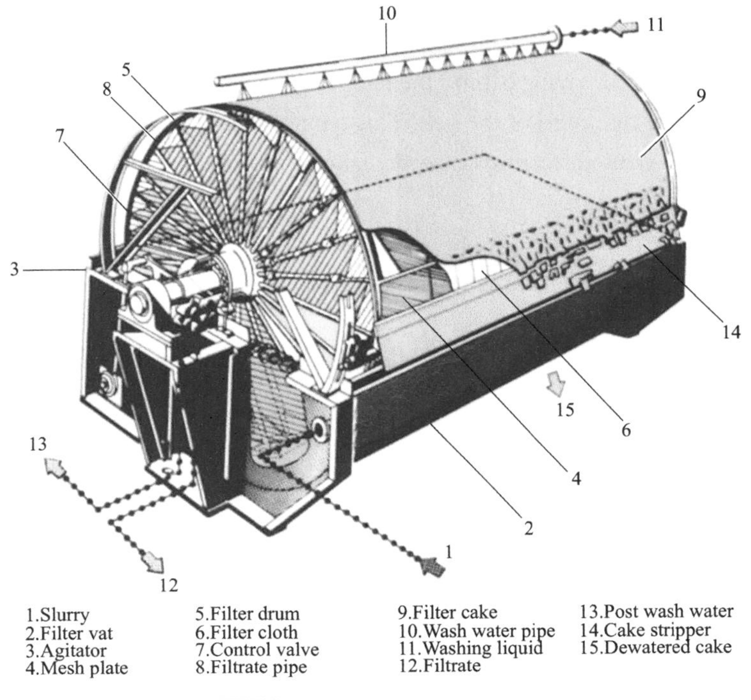

Figure 10.10 Diagram of a rotary drum filter

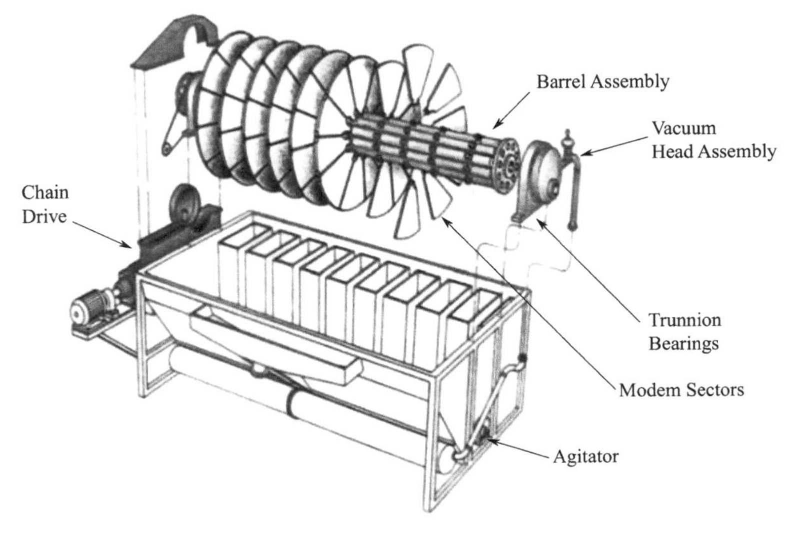

Figure 10.11 Diagram of a rotary vacuum disc filters

The rotary horizontal filters consist of a lotted or perforated endless belt supporting the filter cloth which is transported over the vacuum or suction box (underneath the filter cloth in Figure 10.12). The slurry is fed at one end (feed box at left end) and the filtrate is drawn through the vacuum box, leaving behind the filter cake which is then dewatered further before it is discharged at the right end of the belt. This type of filters has a low capital cost per unit area and can handle diluted slurries; however, cake moistures are higher than that with drum filters.

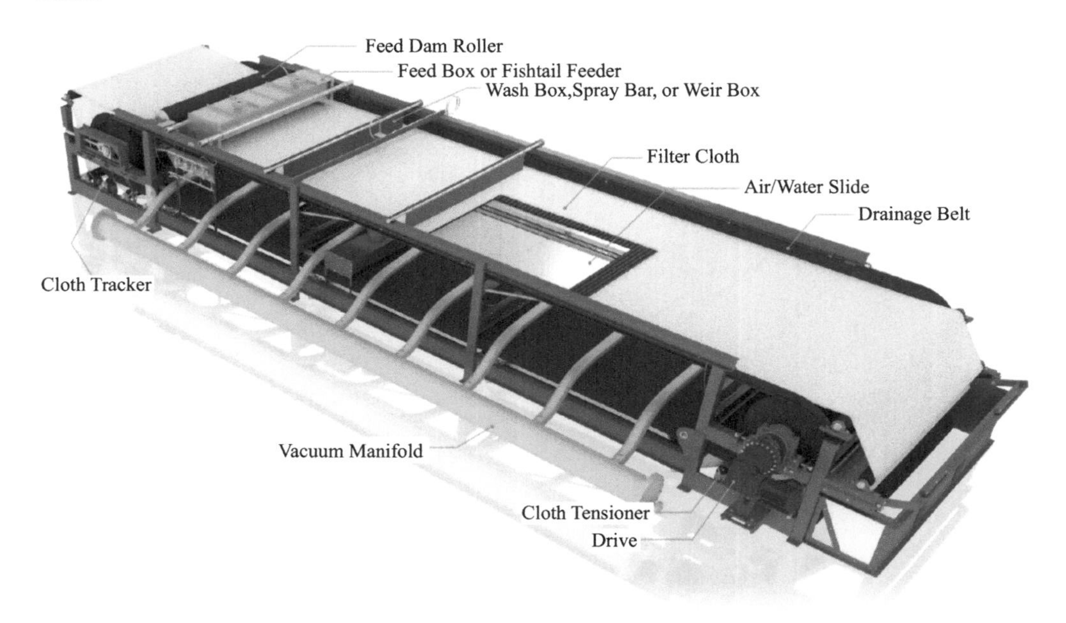

Figure 10.12 Horizontal belt filter (WesTech Engineering, LLC)

10.7 Fluidized-Bed Coal Thermal Dryer

Figure 10.13 shows a coal thermal dryer system. Dewatering of coal is done by bringing together hot gases and wet coal on a continuous "gas flow-coal feed" basis. The drying side of the system operates under negative pressure and the gas scrubbing side operates on positive pressure, both induced by the exhaust fan placed between them.

Hot gases are drawn from the furnace (item 5) and mixed with ambient air through the tempering damper entering the dryer (item 7), under the bedplate, at 1,200°F (648.9℃) usually. Gas flow through the bedplate is such that the coal flowing over it is "fluidized" or floated. The hot gases evaporate the surface moisture which is drawn through the cyclones (item 9), and the dewatered coal, or product, is discharged to a belt conveyer (item 8).

The cyclones discharge the largest air born particles through air locks at the bottom, and moisture laden gases with fine particles to the exhaust fan (item 11). The exhaust fan creates the negative pressure on the dryer side by drawing gases through the system and pushing them through the scrubber side, thus pressurizing it.

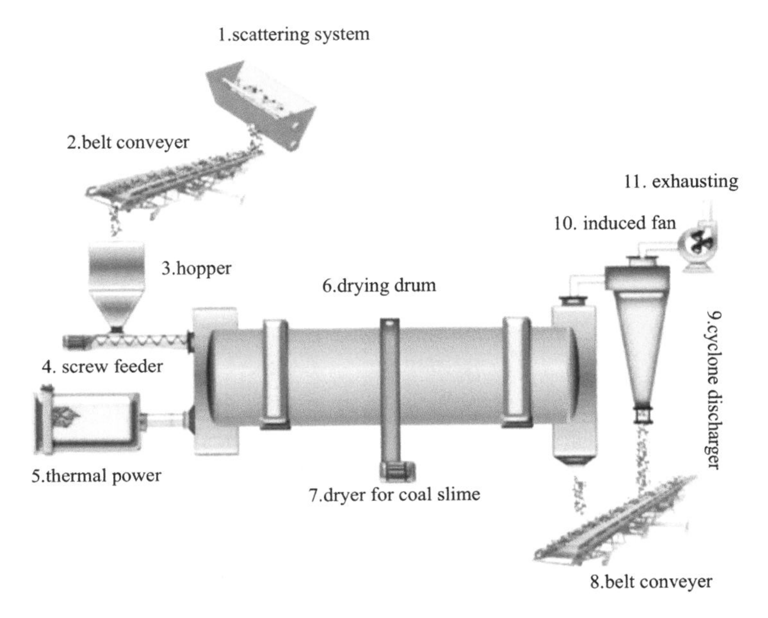

Figure 10.13 Coal thermal dryer system (ABC Machinery)

10.8 Flocculation and Coagulation

Chemical additives are often needed to increase the efficiency or throughput of a dewatering systems. The use of chemical additive must be carefully considered because they are expensive and their use will increase the processing costs. A proper application of chemical additives will have the following advantages:

① Increase separation efficiency,

② Increase throughput, and

③ Require a minimum investment for mixing equipment.

There are, however, several disadvantages to the use of chemical additives:

① They can be expensive,

② A large concentration of some additive may be needed to produce the required results, and

③ Not all slurries are responsive to chemical additives.

The commonly used polymers in coal and mineral industries as flocculating agents are acrylamide homopolymer as well as acrylamide and sodium acrylate copolymer. They are made of free radical polymerization as shown below:

(1) acrylamide homopolymer

$$\underset{\underset{NH_2}{|}}{\left(CH_2-\underset{\underset{C=O}{|}}{CH}\right)_X}$$

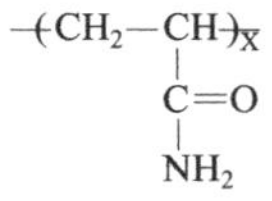

(2) copolymer of acrylamide and sodium acrylate

$$\left(CH_2-\underset{\underset{\underset{NH_2}{|}}{C=O}}{\underset{|}{CH}}\right)_X \left(CH_2-\underset{\underset{\underset{O^-Na^+}{|}}{C=O}}{\underset{|}{CH}}\right)_Y$$

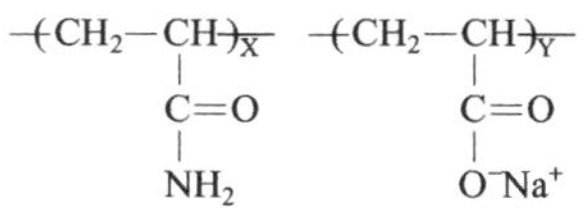

Most polymers used in mineral beneficiation applications are water soluble and differ in performance because of differences in the chemistry of the pendant functional groups. The functional water-soluble polymers are broadly classified into three categories, nonionic, anionic, and cationic, as determined by the chemistry of the pendant groups. They are:

(1) Nonionic

polyacrylamide, X up to 150,000:

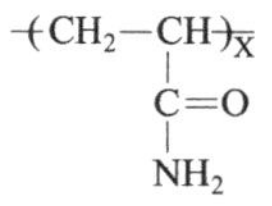

$$\left(CH_2-\underset{\underset{\underset{NH_2}{|}}{C=O}}{\underset{|}{CH}}\right)_X$$

(2) Anionic

sodium polyacrylate, X up to 150,000:

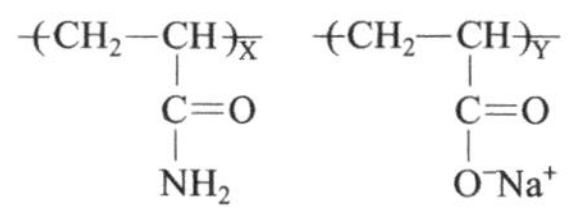

$$\left(CH_2-\underset{\underset{\underset{NH_2}{|}}{C=O}}{\underset{|}{CH}}\right)_X \left(CH_2-\underset{\underset{\underset{O^-Na^+}{|}}{C=O}}{\underset{|}{CH}}\right)_Y$$

(3) Cationic

poly-DADMAC, X up to 600

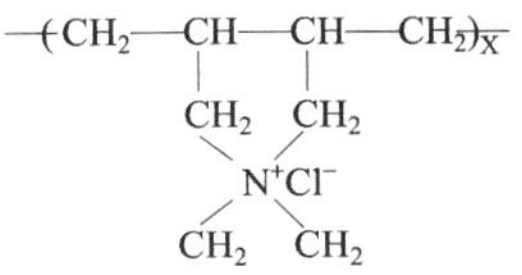

$$-\!\!\left(CH_2-\underset{\underset{\displaystyle CH_2\quad CH_2}{\diagdown\; N^{+}Cl^{-}\;\diagup}}{CH-CH}-CH_2\right)_X-$$

(with CH_2 groups bonded to $N^{+}Cl^{-}$, which carries two further CH_2 groups)

Table 10.2 lists the polymer usage in coal and mineral beneficiation. They can be in the form of solution, emulsion, power, or flocculant.

Solution – Polymer is solubilized in water made available as a pumpable liquid. Polymer concentration of the solution is limited by solubility and solution viscosity.

Emulsion – Many high molecular weight polymers used as flocculants are prepared as oil-continuous inverse emulsions. This form allows for delivery of a high concentration liquid which solubilizes rapidly upon dilution in water.

Powder – Polymers used as flocculants are often made available in a granular or pulverized dry form. Dissolution is mechanically somewhat more intensive as compared to liquid forms, but powders can represent a very cost-effective form for use.

Flocculant – Most mineral particulates are negatively charged in the various aqueous processing environments. The most commonly employed polymeric flocculants are nonionic or moderate anionic. The flocculants used in mining industry act as a bridge mechanism, whereby the polymer molecules are adsorbed onto two or more particles, physically holding them together.

The primary roles of the functional groups including acrylamide groups and carboxylate groups present in the polymer in most mining applications have the capacity to absorb onto mineral particles. The primary function of the acrylamide group in copolymers is to build strong floc by hydrogen bonding. While that for the carboxylate groups is to extend the polymer chain in solution by electrostatic repulsion, enabling bridging to take place more easily.

Table 10.2 Polymer used in coal and mineral beneficiation

Polymer	Polymer Chemistry	Available Forms
Flocculant	Polyacrylamide (nonionic)	Dry; Emulsion
	Acrylamide / Sodium Acrylate Copolymers (Anionic)	Dry; Emulsion
	Sodium or Ammonium Polyacrylates (Anionic)	Dry; Solution; Emulsion
	Acrylamide/DMAEM or DMAEA QUAT	Dry; Emulsion
	Copolymers (Cationic)	Dry;
	Polyethylene Oxide (Nonionic	Suspension
	Guar Gum	Dry
	Hydrolyzed Starch	Dry
	Acrylamide / Sodium 2-AMPS Copolymers	Emulsion
	(Sulfonate Anionics)	Dry; Solution
	Modified Polysaccharides	

Polymer	Polymer Chemistry	Available Forms
Coagulant	Poly DADMAX DADMAC / Acrylamide Copolymers Epichlorohydrin / Dimethylamine Condensation Polymers Poly Aluminum Chloride Amine / Formaldehyde Condensation Resins	Dry; Solution Solution; Emulsion Solution Solution Solution
Dispersant	Polyacrylic Acid, Sodium or Ammonium Salts (Low MW) Methyl acrylate / Acrylic Copolymers and Salts (Low MW) Acrylate / 2-AMPS Copolymers (Low MW) Methacrylate / Acrylic Copolymers and Salts (Low MW)	Dry; Solution Solution Solution Solution
Antiscalant	Polyacrylic Acids and Salts (Low MW) Polymaleic Acid and Salts (Low MW) Polyacrylate / Organo – Phosphonate Blends Polyacrylate / 2-AMPS Copolymers (Blends) Sulfoakylated Polyacrylamides (Low MW)	Solution Solution Solution Solution Solution
Dust Control Binders	Polyvinyl Acetate Vinyl Acetate / Acrylic Copolymers Styrene / Butadiene Copolymer Resins Modified Polyacrylamides	Emulsion Emulsion Emulsion Emulsion

10.9 Sedimentation, Thickening, and Clarifying

10.9.1 Sedimentation Rate and Thickener Capacity

(1) Conventional thickening

Thickening and clarifying are used as solids/liquid separation process in coal preparation plant. For conventional continuous thickeners, feed slurry enters the thickener from the top or bottom of the tank. The particles in feed settle down to the bottom. The supernatant overflows out and the sediment is gathered to the center of the tank bottom by a mechanical device, i.e., rake, as the underflow, where it is then pumped out. The operation for a continuous thickener proceeds by continuously being fed, overflowing the clarified water, and discharging the underflow. There are four zones in the conventional thickener, from top to bottom, the supernatant (A), free-settling zone (B), the zone of compression (C) and disturbed area as given in Figure 10.14. Free settling velocity can be calculated according to Stoke's equation for non-spherical particles:

$$v_{rf} = \frac{g(\rho_s - \rho_l)(\frac{D_p}{\lambda})^2}{18\mu} \quad (10\text{-}27)$$

where v_{rf} is free settling velocity of a particle, ρ_s and ρ_l stand for the densities of solids and liquid, respectively, μ is the viscosity of liquid, g is gravity acceleration, D_p represents the diameter of the particle, and λ is a shape factor for non-spherical particles. For coal slurry or coal refuse, λ is about 1.4 - 1.5.

When slurry concentration is higher than 1~2 percent, particles settle in hindered state. The hindered settling velocity of particles can be calculated by using the expression:

$$v_{rh} = v_{rf}(1-\delta)^m \quad (10\text{-}28)$$

where v_{rh} and v_{rf} are hindered and free settling velocities of particles, respectively, m is experimental constant, δ is the concentration of solids in feed, % by volume, $\delta = 1/(F+1)$, F is the solid to liquid ratio in feed by weight.

(2) Coe-clevenger method

In particle settling, the cross-sectional area of thickener is a key parameter. Coe and Clevenger (1916) presented the following equation to calculate the settling area:

$$A = 1.33\frac{(F-D)}{V_{rf}} \quad (10\text{-}29)$$

where F is the solid to liquid ratio in feed by weight, D is the solid to liquid ratio in underflow by weight, V_{rf} is the free settling velocity of the particles, A is the thickener area (ft^2) required to settle one ton (short ton) dry solids per day (24 hours). In the thickener, there are various layers of different consistency between that of the feed and that of underflow. Therefore, tests must be carried out to obtain the settling velocities of particles under different solid to liquid ratio in feed (the dilution conditions) to determine the maximum settling area requirement by Equation (10-29).

Example: Thickener design based on the Coe-Clevenger method

A coal preparation plant has a water circuit that operates at 500 gpm of coal fine slurry with a solid to liquid ratio of 1∶10. In considering the use of a thickener to prevent slime buildup, tests on slurry settling were conducted to determine its settling rate. The Coe-Clevenger method was employed to determine the required settling area of the thickener. The specific gravity of the solids (silica minerals) is 2.8. The final suspension density was 1∶1 (50% wt solids in underflow). The test results are shown in Table 10.3.

Table 10.3 Results of batch solid settling tests

Solid to Liquid Ratio	Settling Rate (ft/hr)
1 : 9	2.00
1 : 3	0.30
1 : 2	0.15
1 : 1.5	0.08

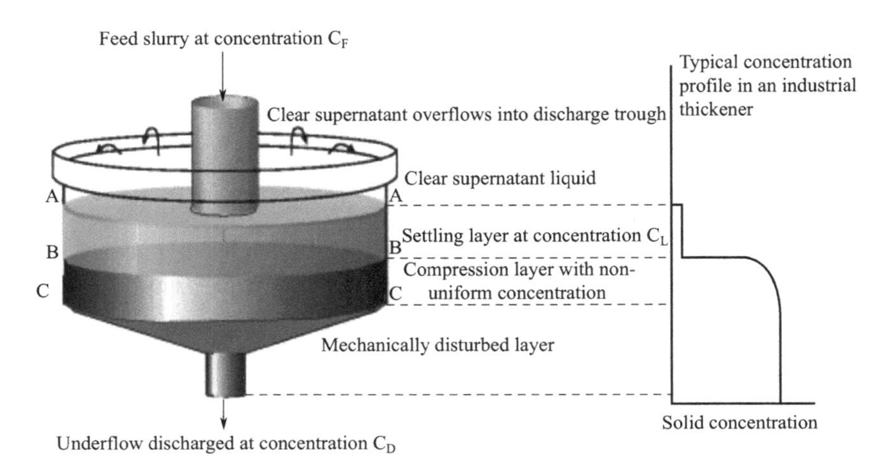

Figure 10.14 Schematic representation of the ideal thickener operating at steady state

Solution:

The estimated unit areas (ft^2) per ton of dry solids per 24 hrs are:

$$F=\frac{9}{1}\quad D=\frac{1}{1},\ A=\frac{1.333\times(9-1)}{2.0}=5.33$$

$$F=\frac{3}{1}\quad D=\frac{1}{1},\ A=\frac{1.333\times(3-1)}{0.3}=8.89$$

$$F=\frac{2}{1}\quad D=\frac{1}{1},\ A=\frac{1.333\times(2-1)}{0.15}=8.89$$

$$F=\frac{1.5}{1}\quad D=\frac{1}{1},\ A=\frac{1.333\times(1.5-1)}{0.08}=8.33$$

$$500\ \text{gpm}\times\frac{3785\ \text{ml}}{1\text{gal}}\times\frac{60\ \text{min}}{1\ \text{h}}\times\frac{24\ \text{h}}{1\ \text{day}}=2.7252\times10^9\ \text{ml / day}$$

Let the weights of solid and water be x and y grams, respectively.

$$\frac{x}{y}=\frac{1}{10}=0.1$$

$$\frac{x}{2.8}+\frac{y}{1}=2.7252\times10^{9}$$

Solve the simultaneous equations shown above, yield $x = 2.6312\times10^{8}$grams

$$2.6312\times10^{8}\frac{g}{day}\times\frac{1lb}{454g}\times\frac{1ton}{2000lb}=289.78\ ton/day$$

Therefore, the maximum thickener area is:

$A = 289.78\ tons \times 8.89 \times (1+0.25)\ ft^2 = 3220\ ft^2$ (note: safety factor is 25%),

since $A = \pi(\frac{D}{2})^2$; $D=\sqrt{\frac{4A}{\pi}}$

Diameter of thickener $=\sqrt{\frac{4\times3220}{\pi}} = 64.03 \sim 65$ ft

10.9.2 Kynch Method

Kynch presented his thickening theory in 1952 (Kynch, 1952). It states that the settling velocity of particles is a function of local concentration, i.e., $v = f(C)$. In a continuous thickener operation, due to continuous overflow, there is an upward flow in the tank. If the velocity of upward flow is u, the relative settling velocity of particles is C dV/dC, then the absolute settling velocity of particles can be expressed as

$$V_t = V\frac{dV}{dC} - u \tag{10-30}$$

It clearly shows that the hindered settling velocity of particles must be larger than the upward flow velocity so that the solid particles can settle. Figure 10.15 shows a typical batch settling curve. With the induction in interface height and the increase in slurry concentration, settling velocity of particles decreases. There is a critical concentration C_c, at which the solids flux ($G_i=C_iv_i$) reaches a minimum value G_{min} as shown in Figure 10.16. This minimum value of flux limits the capacity of thickening. The thickening area required can be calculated according to this value. If no solid is present in the overflow and the underflow concentration C_u equals to critical concentration C_c, then

$$C_f = \frac{C_c Q_u}{Q_f} = \frac{C_u Q_u}{Q_f} \tag{10-31}$$

The mass balance for solid is

$$Q_o = Q_f(1-C_f) - Q_u(1-C_u) \tag{10-32}$$

The overflow (upward flow) velocity is

$$u = \frac{Q_o}{A} = \frac{Q_u C_u}{A}\left(\frac{1}{C_f} - \frac{1}{C_u}\right) \tag{10-33}$$

Let the hindered settling velocity of the largest particle present in the overflow is equal to the overflow velocity u, the minimum settling area should be:

$$A_{min} = \frac{Q_u C_u}{V_t}\left(\frac{1}{C_f} - \frac{1}{C_u}\right) = Q_u C_u \left[\frac{\frac{1}{c} - \frac{1}{c_u}}{V_t}\right] \tag{10-34}$$

where C_f is feed concentration, Q_u is underflow volume, Q_f stands for feed volume, and Q_o is overflow volume.

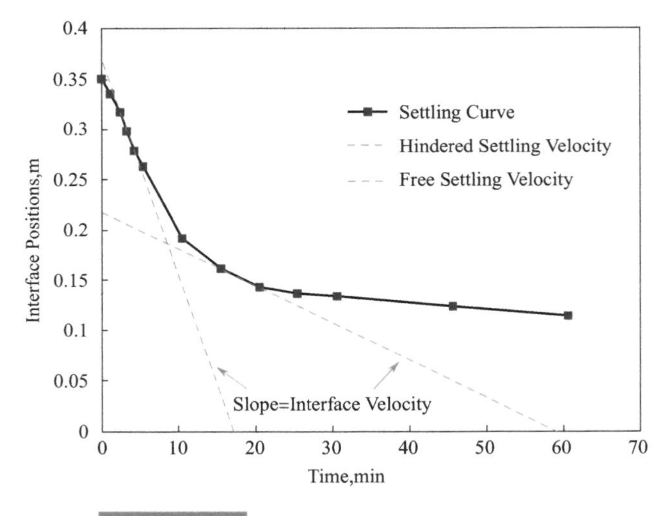

Figure 10.15 A typical batch settling curve

Example:

The feed rate of a thickener is 500 m^3/hr, and the required solids in the feed in order to produce 5 percent solids underflow density is 35 percent. The settling test shows that the settling velocity is 24 m/hr. Find the required settling area.

Solution:

$$Q_u = \frac{Q_f C_f}{C_u} = \frac{500\left(\frac{m^3}{h}\right) \times 5\%}{35\%} = 71.43\left(\frac{m^3}{h}\right)$$

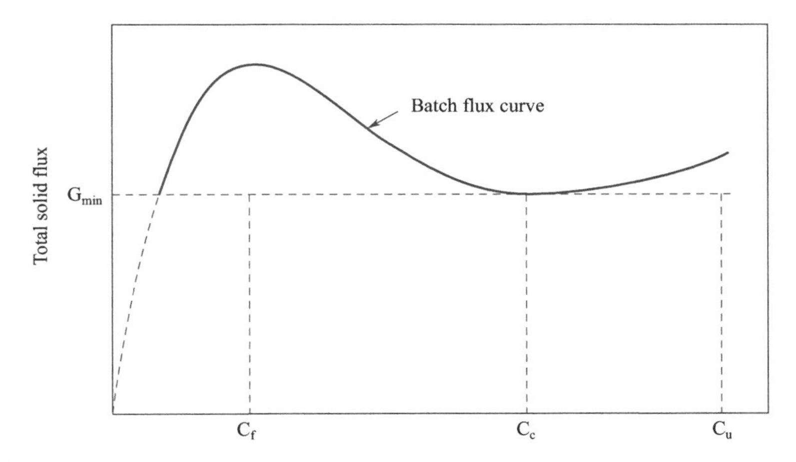

Figure 10.16 Solid flux as a function of slurry concentration

$$A_{min} = \frac{71.43\left(\frac{m^3}{h}\right) \times 35\%}{24\left(\frac{m}{h}\right)} \times \left(\frac{1}{\frac{5}{100}} - \frac{1}{\frac{35}{100}}\right) = 17.85(m)$$

Equation (10-29) is often used for thickener design when a non-flocculated slurry is involved. Equation (10-34) is usually used for the thickener dealing with flocculated slurries. Both equations indicate that thickener capacity is related to the settling velocity of particles and the underflow concentration. When underflow concentration is specified, the settling velocity of particles is the key factor affecting the thickening capacity. From Equations (10-27) and (10-28), for a given slurry, the settling velocity of a particle is directly proportional to the diameter square of the particle. Therefore, to enhance the settling rate, it must increase the particle size by formation of floc using flocculants.

10.9.3 Operating Line

For various initial slurry concentrations, a group of settling curves are obtained by using the batch cylinder settling rate tests. The initial settling rates can be determined by taking the slope of the settling curve as shown in Figure 10.15. Then the settling rate and slurry concentration are described by a straight line using the semi-log coordinates as shown in Figure 10.17. The transformed data are represented by

$$\log v = mC + b \tag{10-35}$$

or

$$v = 10^{(mC+b)} \tag{10-36}$$

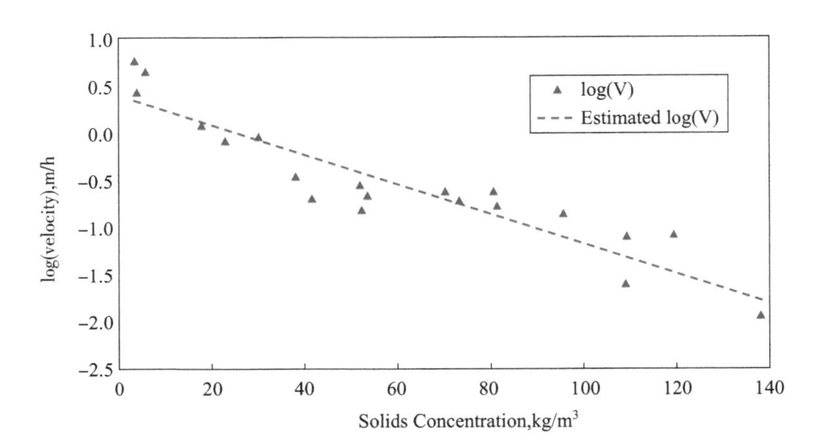

Figure 10.17 Semi-log plot of data produces a straight line

where C is the slurry initial concentration, m and b are the fitting or regression parameters. Using the fitted data, the settling flux curve may be drawn according to the value $G_i=C_iv_i$ as shown in Figure 10.18.

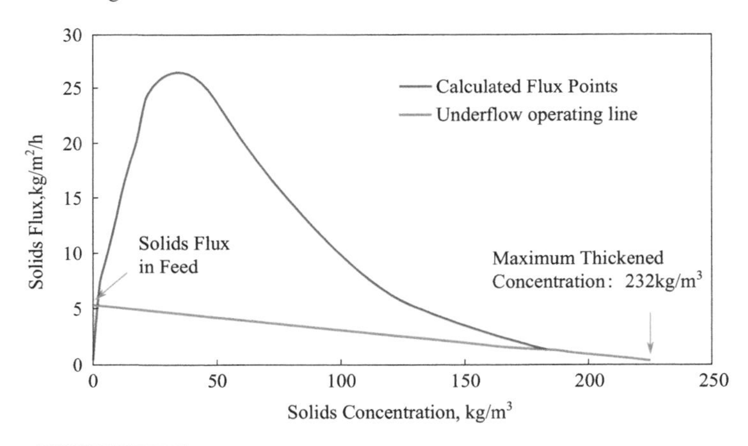

Figure 10.18 Batch flux curve shows the critically loaded unit

When the thickener area is determined, the applied flux is

$$G_f=\frac{C_fQ_f}{A} \tag{10-37}$$

Using G_f, the operating line of a thickener with a certain settling area can be determined by the following method:

① Draw the settling curves of various initial slurry concentrations.

② Obtain the initial settling rates.

③ Find the regression curve for the test data and calculate the flux.

④ Draw concentration-flux curve.

⑤ Construct the operating line tangent to the flux curve at the feed point (flux axis), and intersect the concentration axis. The operating line gives the optimum underflow concentration, and a reasonable underflow pumping rate.

Example:

A thickener has a 15.24m (50ft) diameter, the initial settling rate can be calculated by

$$v = 10^{(-0.0142C+0.370)}$$

What is the optimum underflow pumping rate and the responding underflow concentration, when feed solid is 177,600 kg/hr and feed concentration is 5 kg/m^3, and feed density is 1,000 kg/m^3?

Solution:

Using the calculated data to draw the flux curve as shown in Figure 10.18.

$$A = \pi r^2 = \pi \times 15.24^2 = 182.4\left(m^2\right)$$

$$G_f = \frac{\left(C_t Q_f\right)}{A} = \frac{\left(5 \times 177{,}600\right)}{\left(1000 \times 182.4\right)} = 4.87\left(\frac{\frac{kg}{m^2}}{h}\right)$$

The operating line is tangent to the curve at and starts from 4.87 (flux axis) and intersects the concentration axis at 232 kg/m^3. This is the optimum underflow concentration. The underflow rate can be calculated as

$$Q_u = \frac{F_f Q_f}{C_u} = \frac{\left(5 \times 177{,}600\right)}{232} = 3{,}830\left(\frac{kg}{h}\right)$$

10.9.4 High-capacity thickening

The high-capacity thickening principle is based on the full flocculated mineral slurry system. The main differences between a high-capacity thickener and a conventional thickener are adequate use of flocculant and adoption of deep feed system. For a high-capacity thickener to be effective, the flocculated feed is directly fed to the zone of compression in the tank and diffuses according to a controlled velocity. Because the fully developed flocs have a much higher settling velocity than an individual particle in free settling zone, and using proper feeding and flocculant addition systems, the feed does not have to pass through the supernatant, free settling zone or transition zone. Thus, the settling distance and the settling time are shortened. Some individual particles that fail to be flocculated move up with the upward current may be enmeshed and filtered by the sludge bed and settle down with the sludge. Under this condition, the separating effectiveness is improved, and cleaner overflow water is obtained.

A high-capacity thickener piping line and thickener itself are given in Figures 10.19 and 10.20. Different from the conventional thickener, there are only two zones in the settling tank for the high-capacity thickener: the supernatant and the compression zone. The capacity of high-capacity thickener can be determined by using Equation (10-34). However, the settling velocity of floc should be used to substitute for the settling velocity of the individual particles.

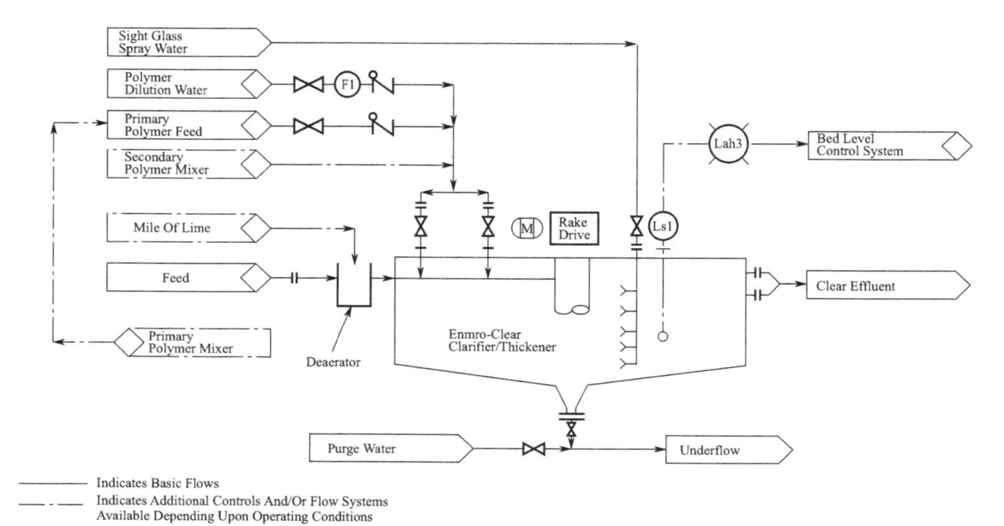

Figure 10.19 High-capacity thickener piping line

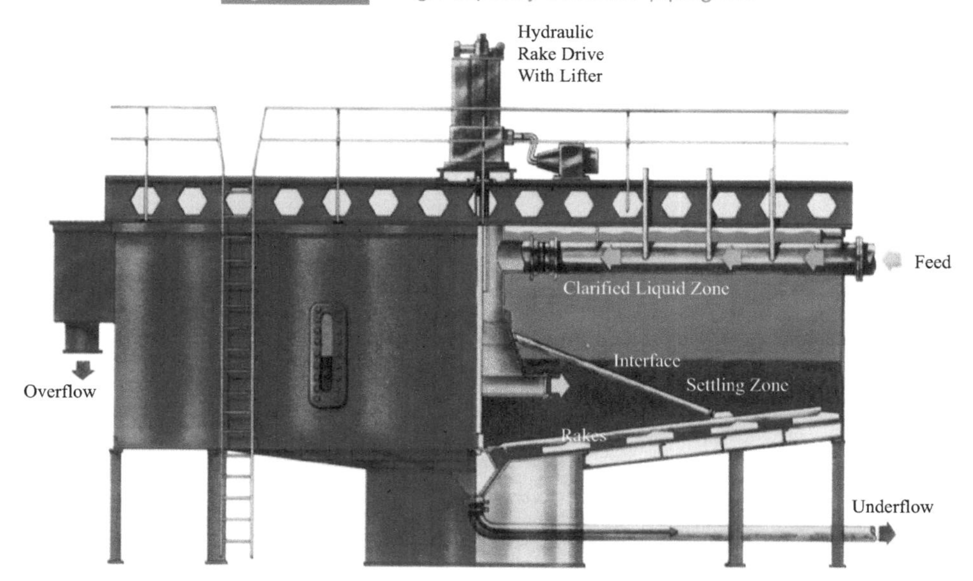

Figure 10.20 Enviro-Clear high-capacity thickener (Enviro-Clear Company, Inc.)

11

ON-LINE COAL QUALITY MONITORING AND MATERIAL HANDLINGS

Determination of coal characteristics is critical to the coal users regarding the feed material quality and production process control. In coal preparation, real time analysis provides valuable information of the properties of the coal product, which allows the operator to make a quick response to the changes in the operating parameters, and thus achieve good stability in coal preparation and favorable homogeneity in coal product quality. As a result, the coal preparation plant is possible to run at an efficient and stable manner.

With the development of technologies in fast analysis of coal quality, there are multiple on-line analyzers available in the market. Various principles of analysis were applied focusing on measuring different aspects of coal characteristics. The measurement of different coal characteristics will assist in the management and utilization of coal in specific downstream applications, such as combustion, reduction, and liquefaction.

Coal online analyzer normally installed on top of belt conveyors to instantaneously monitor the quality of flowing coal streams. A typical online analyzer is composed of three segments - a radiation source, a detector, and an analyzing software. Standard samples are required for calibration of the analyzers periodically. Based on the demand on the analysis on specific assays of mineral, the on-line analyzers can be categorized into different groups that adopt variant technologies to achieve the corresponding accurate measurement.

Application of real time analysis on minerals could disclose the changing conditions of the flowing material, thus allowing efficient process control and efficient quality monitoring. The homogeneity of product quality could be expected to improve with the assistance of on-line analysis technologies. The on-line analytic system generally includes a sampling system for calibration, an analyzer, and a computer software package. According to the desired assay for analysis, the principles employed in the corresponding analyzers are different. Large mines, mineral processing plants, and metallurgical plants could benefit remarkably from the applications of on-line analytic systems, which require sophisticated technical analysis and economic evaluation beforehand as the cornerstone supporting their usage.

11.1 X1-LiNX Coal Analyzer

The SABIA X1-LiNX coal analyzer (Figure 11.1) is a rugged, reliable and high-performance device for installation in mines, coal-prep facilities, power plants, or any facility that requires real-time elemental analysis of coal. Applications include coal seam tracking, sorting, blending, analysis reporting and calibration, process control and much more. Some of the features of this outstanding unit are:

① High performance full elementary analyzer.

② It measures ash, BTU, S, and full elemental analysis.

③ It measures moisture (optional).
④ It is designed for in-line use.
⑤ It has web-based user interface.
⑥ The results are reported over the web pages, and over the PLC interface.
⑦ It is easy to install in existing facilities.
⑧ It is easy to calibrate.

Figure 11.1 Sabia X1-LiNX coal analyzer

11.2 Combined Ash, Moisture and Elementary Belt Analyzer (AM-EBA)

The AM-EBA from ASYS and Energy Technology Inc. (ETI) is an on-line ash, moisture, and elemental analyzer. It combines prompt gamma neutron activation analysis (PGNAA) technology with the electronics and digital signal processing. The AM-EBA system yields immediate benefits to the user such as improved measurement precision, long-term calibration stability, reliability, and ultimate operating efficiency. It is the real-time, belt-load-independent, full stream analyzer which measures moisture, ash, sulfur, nitrogen, chlorine, and coal ash elements with the highest analytical performance. It requires no sampling and scans the entire process coal stream independently of belt load variations. The AM-EBA is used in a wide range of applications for monitoring coal quality and improving productivity and profitability in coal processing. Applications include monitoring, sorting, blending and control of raw coal, preparation plant, and load-out coal feeds.

The model DGA-410 ash meter is a nuclear gauging device for measuring ash weight percentage of coal. The measuring portion of the device, which consists of a source and a detector assembly, is normally mounted across a conveyor belt on existing belt structure. The detector is connected to an electronics enclosure housing an industrial computer which processes the detector signals and displays the measured results to the operator. The ash meter generates ash weight percent and weight/density measurements every three seconds making it useful in online process and control applications. It can control the ETI model HSG1 High Speed Sort Gate for separating process coal into different quality piles.

11.3 Moisture Meter

An on-line moisture meter provides a quick method to gauge mineral moisture content to evaluate operation or monitor product quality. Many types of on-line moisture meter based on various principles have been developed. Near Infrared, radio frequency, microwave, and conductance and capacitance are the prevailing technologies adopted for the moisture meters. The on-line moisture meter has wide application in agriculture, solid fuels, biofuels, building materials, food, and mining.

In coal industry, the users of coal can be super sensitive to the moisture content, which is directly linked to the properties, such as net calorific value, and fixed carbon content, of coal product on as received basis. The values or the fluctuations of these properties influence coal utilization and coal price. In short, the moisture content controls the entire profile of coal product quality. Knowing the moisture content instantaneously would be helpful for the operators to adjust operating parameters of the process that produces or uses coal.

Accurate application of moisture meter can benefit plant productivity, production efficiency, quality control accuracy, dry volume accounting, and health and safety compliance. The on-line moisture meter typically installed on a conveyor belt (Figure 11.2), which enables measurement of moisture content when material passes through the gap between the transmitter and the detector. The moisture meter is easy to install. Advanced

Figure 11.2 On-line moisture analyzer on belt conveyor

software package will allow the data reception and storage, data transmission, human interaction, quick calibration, and remote analysis. The moisture meter is a cost-effective equipment to boost the operating efficiency for many industries.

11.4 Elemental Analyzer

The characteristics of coal such as ash, sulfur, and ash elements are important indicators of coal quality to some users. Knowing these characteristics could assist in differentiating coal product by quality, or even assist in making re-washing decisions on some out-of-spec coal products. Traditional approach to disclose this information is through sampling, sample preparation, and sample analysis. This process easily takes more than 24 hours from the sample intake to the final analytical results.

An online elemental analyzer, as illustrated in Figure 11.3, provides a fast way to unveil the information of coal being transported on the conveyor belt. With the minute-to-minute input from the online elemental analyzer, operators could make quick reaction and take quick actions on process control, such as sorting, splitting, and blending, to ensure the quality of coal product with good homogeneity.

Figure 11.3 On-line elemental analyzer

Gamma neutron activation is a commonly applied technology in the elemental analyzer. The analyzer could be mounted on the conveyor belt where the composition of the total burden of coal could be measured by the analyzer. Some analyzers are equipped with heating value analyzing capability, which normally comes together with the moisture measurement function. Advanced virtual view software is packaged together with the analyzer. Analyzed data are displayed on the interface to the operator. The cross-belt elemental analysis offers an efficient way of quality control enables the operator to make instant decisions to increase plant yield, and thus the bottom line.

11.5 Belt Scale

The belt scale provides an instantaneous reading on the weight of mineral being transported in a conveying belt. The operation, or measurement, of the belt scale is independent of the bulk material being carried. The belt scale can be mounted on any type of belt conveyor, with a heavy or light frame. Due to its convenience in installation and utilization, the belt scale has wild application in different segments of mining industry, including extraction, processing, and storage. Many belt scales are equipped with a self-calibrating function, which offers great convenience to routine inspection and adjustment.

The design of belt scales is divided into three categories. The accuracy of belt scales varies accordingly. The selection of belt scales should be depended on the belt conveyer design, measurement accuracy requirement, and budget allowed.

11.5.1 Singe-Station Conveyor Belt Scale

A single-station conveyor belt scale consists of one roller station, as illustrated in Figure 11.4. A weight trough is incorporated with screws, allowing the height adjustment of the three rollers and the fine tuning of each roller, which must be aligned with both the upstream and downstream stations. The typical trough angle ranges from 20° to 45°. Two standard weighting brackets, normally designed outside of the conveyor, are equipped to suspend the standard weights and carry out calibration. Several strain gauge sensors, overload stop, and clamping system are generally included in the single-station conveyor belt scale. The accuracy of single-station conveyor belt ranges from ±0.5% to ±1%.

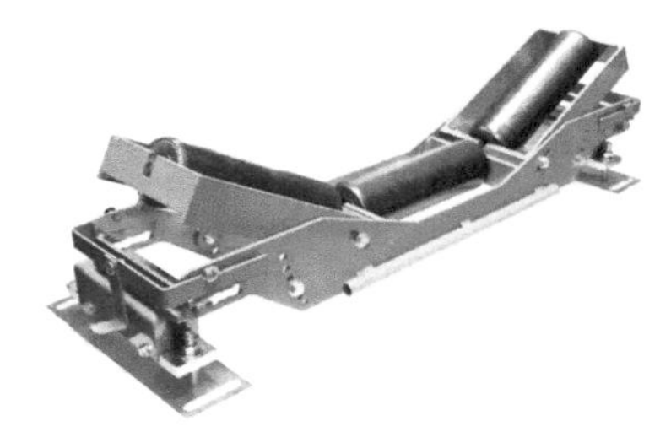

Figure 11.4 Single-station Belt Conveyor Scale

11.5.2 Dual-Station Conveyor Belt Scale

A dual-station conveyor belt scale consists of two single-station conveyor belt scales installed in series (Figure 11.5). The space between the single stations is adjustable according to the installation requirements. The combination of two belt scales uses parallelogram style load cells that provide fast reaction to vertical forces and instantaneous response to product loading. The dual-station conveyor belt scale is capable of monitoring fast moving belts for continuous in-line weighing. Additionally, it is more suitable for uneven or light product loading. As such, it allows the measurement of vast variety of products, even powders, in all sectors of mining industry. Exceptional accuracy, up to 0.25%, and good repeatability could be expected from the dual station designed conveyor belt scale.

Figure 11.5 Dual-station conveyor belt scale

11.5.3 Multi-Station Conveyor Belt Scale

A multi-station conveyor belt scale, as illustrated in Figure 11.6, consists of more than two roller stations. One suspended chassis consists of two side rails and more than two trough roller stations forms the basis of the weighting table of a multi-station conveyor belt scale. Equipped with multiple gauge sensors, the assembled weighting table assists in alleviating the mechanical constraints and banding that disturb the weight measurement. The whole weighting table is stabilized by multiple ties to the fixed frame, which also allows the inclining installation of the belt scale. The height adjustment of the rollers and their adjustment with respect to the upstream and downstream stations are achieved by several screws built in each weighing station. Weight brackets are applied in the weighing table to achieve dynamic calibration. The expected accuracy, up to ±0.125%, is higher than the single or dual station conveyor belt scale.

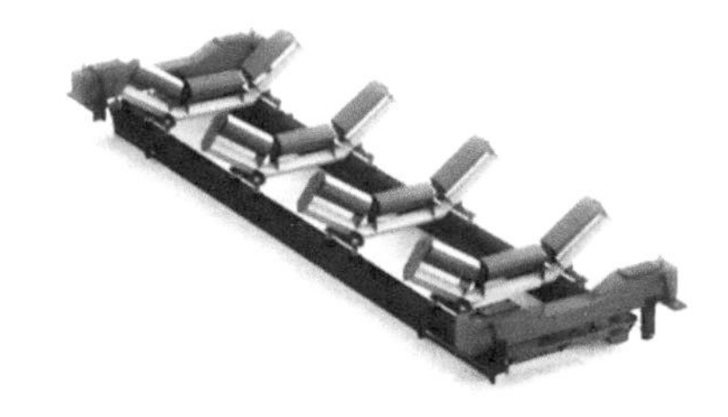

Figure 11.6 Multi-station conveyor belt scale

11.6 High Speed Sorting Gate

The high-speed sorting gate is a rugged gate assembled at the end of a conveyor belt (Figure 11.7). Materials could be separated by the high-speed sorting gate by quality discrepancy. It is equipped to measure the quality of flowing material, which is then split according to the preset point of product quality by the operator. In general, the control of sorting can be automatic or manual.

The sorting function could differentiate the ore by grade. When it is applied in the coal industry, the measurement of ash content is used to divide the run-of-mine coal, according to the quality requirements of the market, into various products, thereby reducing the amount of low-ash coal waste and improving the operating profit margin. The sorting gate typically has no limit on flow capacity. It is a valuable instrument to the mines that produce products with distinct variance in the quality of different components.

The high-speed sort gate can be applied to process control. For example, the sort gate could be installed at the outlet of a coal preparation plant. In this case, the sort gate could not only secure the product quality, but also form a closed-loop in conjunction with the dense medium controlling system. Using this closed-loop design, the coal preparation plant could more homogeneously produce coal product with 100% assurance of the product quality. In another instance, the high-speed sort gate could be installed on the feed belt of a coal power plant. The sort gate could timely remove significant contaminants from the feed coal. Additionally, the sort gate would send signals to the boiler to make ready adjustments reducing boiler fouling and slagging, and consistently improving the heat rate.

Figure 11.7 High speed sorting gate

PLANT FLOWSHEET DESIGN

A complete full-fledged coal preparation plant includes from raw coal stockpiles through cleaning plant to clean coal loadout facilities.

In raw coal storage, it may consist of:

(1) Raw coal storage silos

① Traversing belt feeder;

② Vibrating feeder.

(2) Conical stockpiles

① Stacking tube;

② Reciprocating feeders.

12.1 Coal Cleaning Circuits Configuration and Plant Flowsheet Design

As shown in Figure 12.1, a complete coal cleaning plant consists of coarse, intermediate, fine and ultrafine circuits by size interval, and sizing, concentration, and de-watering by operation.

The first step in coal cleaning is sizing, in which raw coal is crushed by the following one or by combination of several or all devices:

① Grizzly scalping screen;

② Rotary breaker (HGI up to 62);

③ Double roll crusher (HGI up to 100).

After sizing, raw coal of various sizes is fed to four separate circuits for cleaning according to their size range, i.e., coarse, intermediate, fine and ultrafine cleaning circuits. The equipment used in each circuit may consist all or parts of the lists under each cleaning circuit as follows:

(1) Coarse Coal Cleaning Circuit: dense media vessel

① Raw coal screen;

② Pre-wet screen;

③ Dense media vessel, or water-based Baum jig;

④ Fixed sieve, drain and rinse screens for clean coal product and refuse;

⑤ Coarse clean coal (coarser size fraction) crusher;

⑥ Coarse clean coal (finer size fraction) crusher;

⑦ Refuse bin and truck (or conveyor belt);

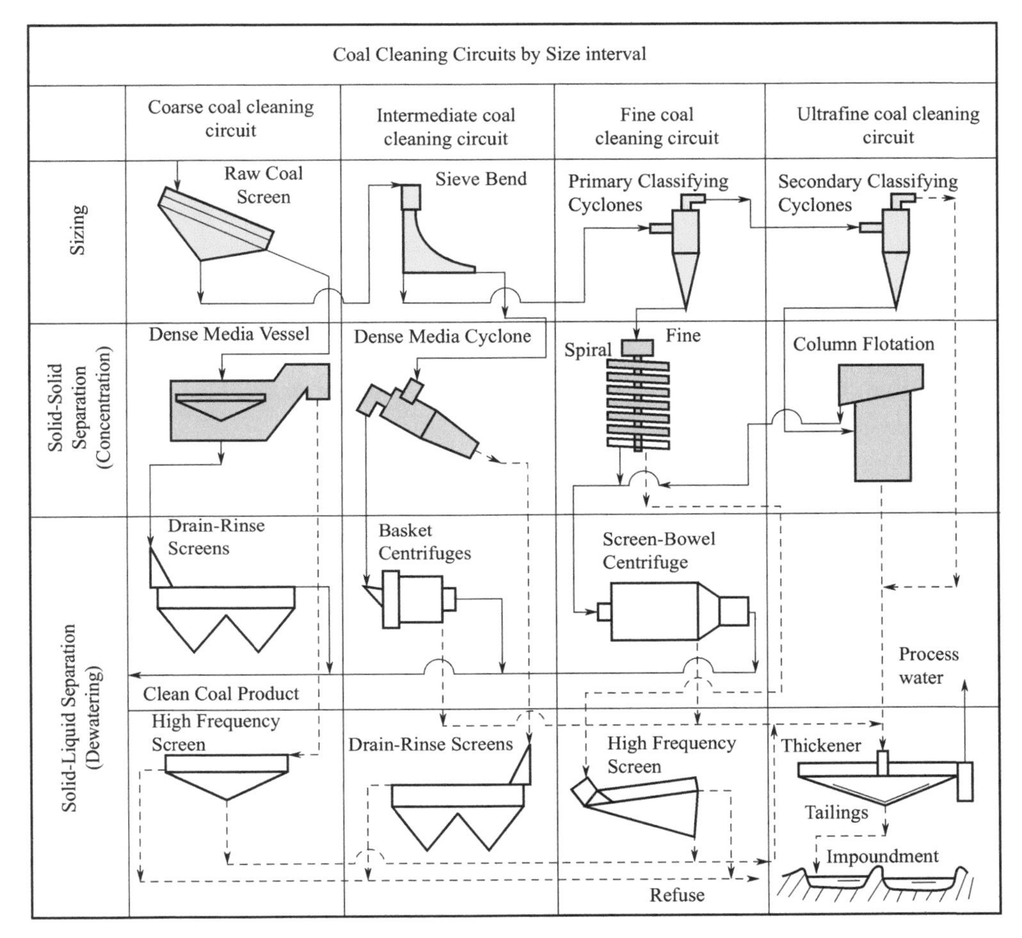

Figure 12.1 Coal cleaning circuits by size interval

⑧ Dense media sump and pump;

⑨ Dilute sump and pump;

⑩ Magnetic separator;

⑪ Magnetite storage bin and feeder.

(2) Intermediate Coal Cleaning Circuit: dense media cyclone

① Sieve bend;

② Dense media cyclone, or water only cyclone, water-based concentration table;

③ Fixed sieve, drain and rinse screens for clean coal product and refuse;

④ Clean coal product centrifugal dryer;

⑤ Refuse bin and truck (or conveyor belt);
⑥ Dense media sump and pump;
⑦ Dilute media sump and pump;
⑧ Magnetic separator;
⑨ Magnetite storage bin and feeder.

(3) Fine Coal Cleaning Circuit: spiral concentrator or teeter bed separator (TBS)

① Classifying cyclone;
② Spiral concentrator or TBS (hindered-settling bed separator);
③ Primary classifying cyclone;
④ Primary classifying cyclone sump and pump;
⑤ Vibrating screens for the clean coal product and refuse;
⑥ Clean coal product centrifugal dryer;
⑦ Clean coal product screen bowl;
⑧ Refuse bin and truck or conveyor belt.

(4) Ultrafine Coal Cleaning Circuit: flotation tank cell or flotation column

① Classifying cyclone;
② Flotation tank cells or flotation column;
③ Secondary classifying cyclone;
④ Secondary classifying cyclone sump and pump;
⑤ Fine refuse vibrating screen;
⑥ Clean coal product centrifugal dryer;
⑦ Clean coal product screen bowl;
⑧ Refuse bin and truck (or conveyor belt).

Figures 12.2 – 12.9 show examples of flowsheets for different segments or cleaning devices, while Figure 12.10 shows the detailed flow sheet for a 6-million clean ton annual production coal preparation plant.

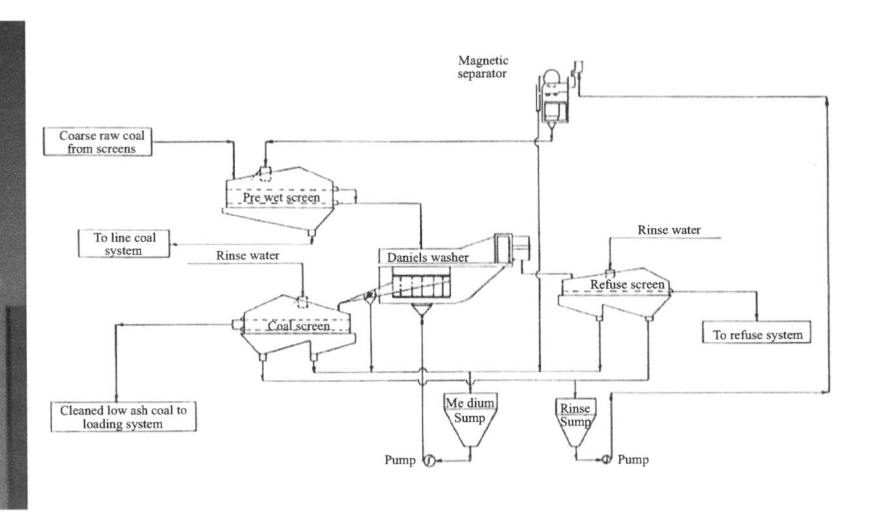

Figure 12.2 Coarse coal cleaning circuit using dense media vessel (drag type)

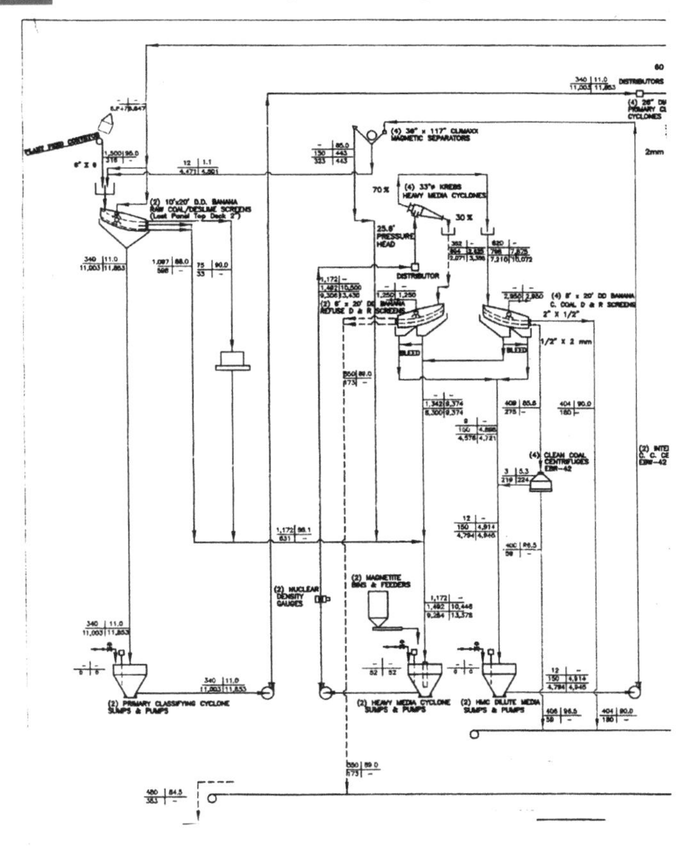

Figure 12.3 Coarse or intermediate coal cleaning circuits using dense media cyclones circuit

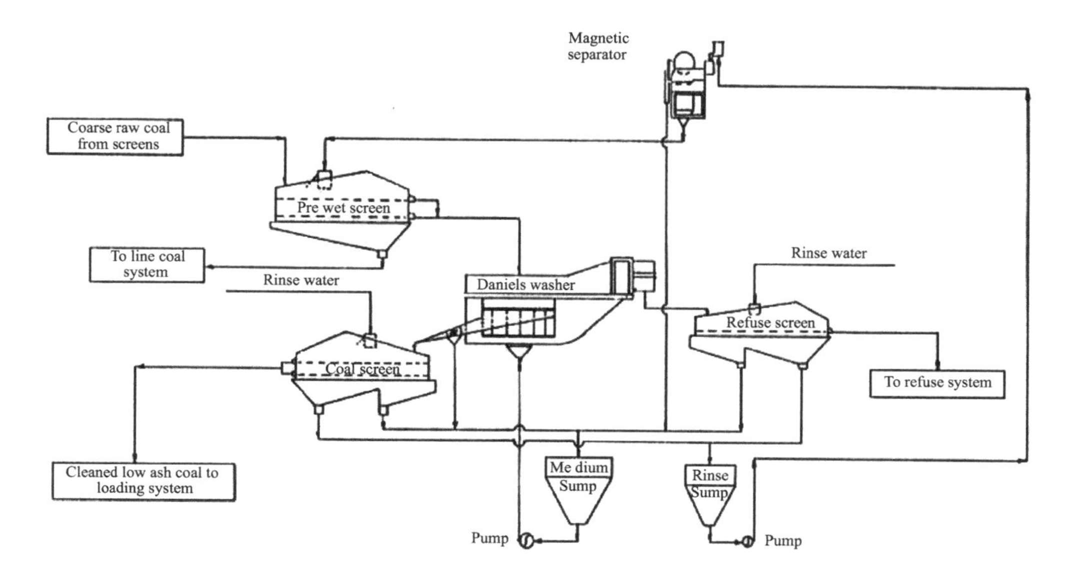

Figure 12.4 Dense media vessel circuit

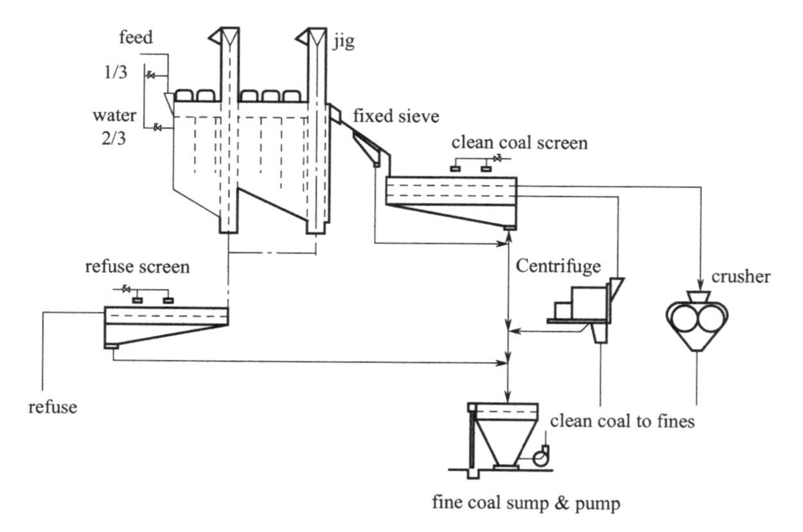

Figure 12.5 Jig circuit

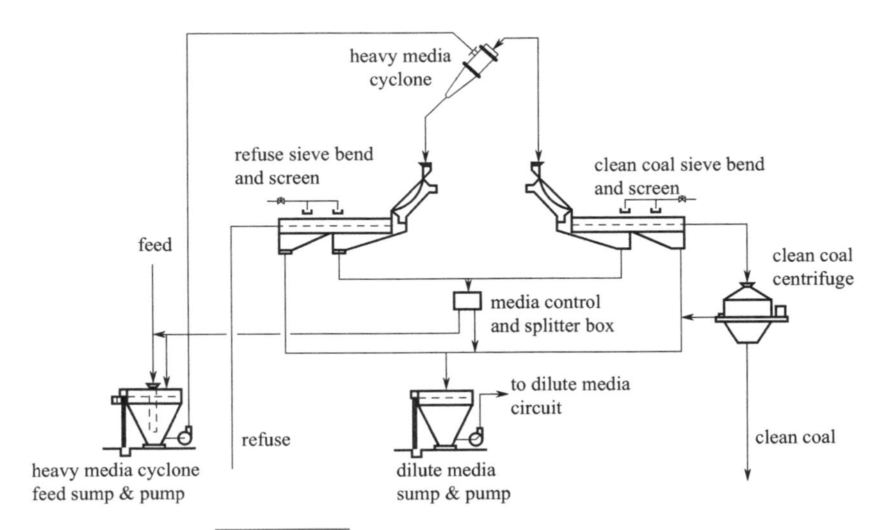

Figure 12.6 Heavy media cyclone circuit

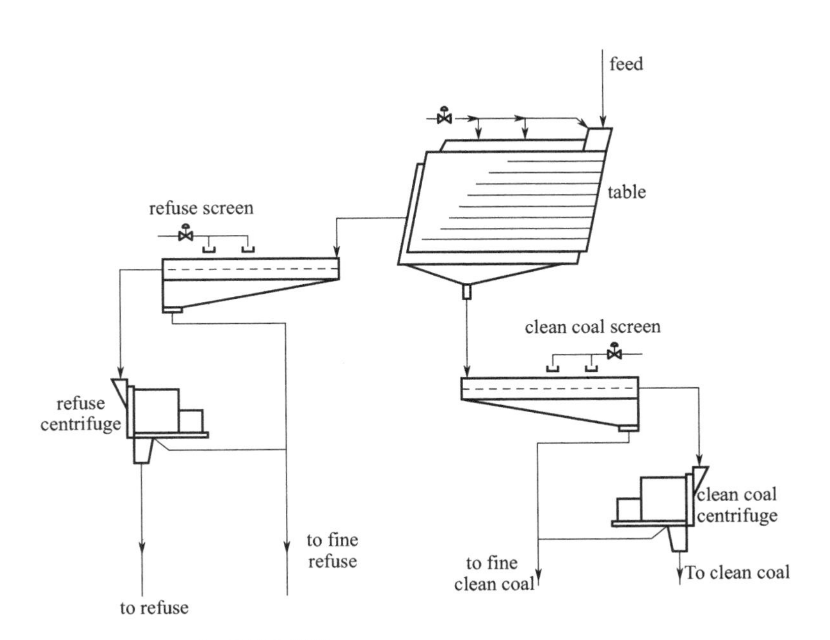

Figure 12.7 Concentration table circuit

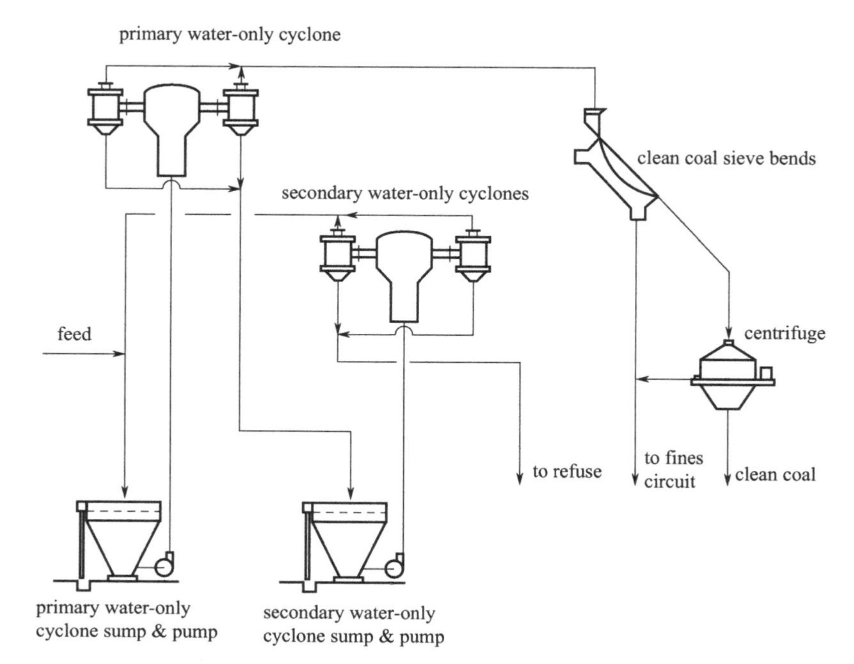

Figure 12.8 Hydrocyclone circuit

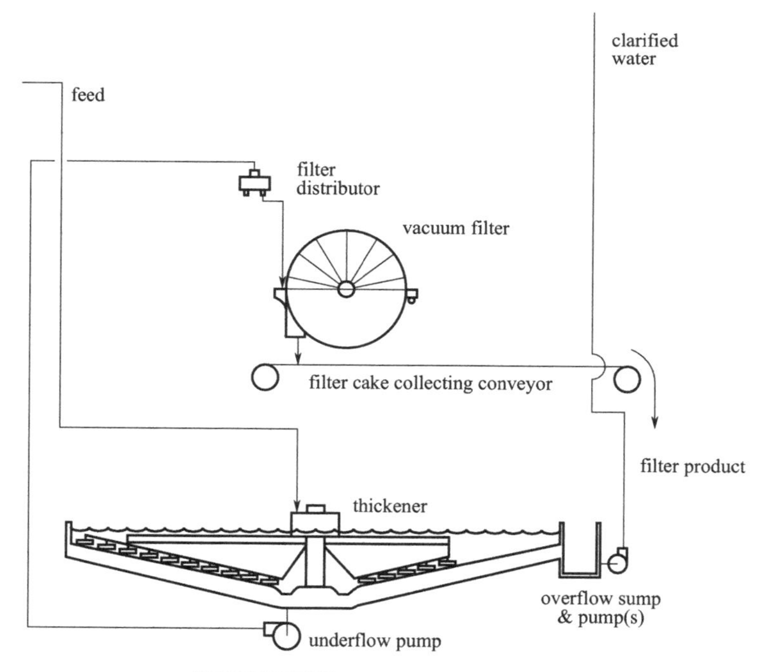

Figure 12.9 Refuse/water circuit

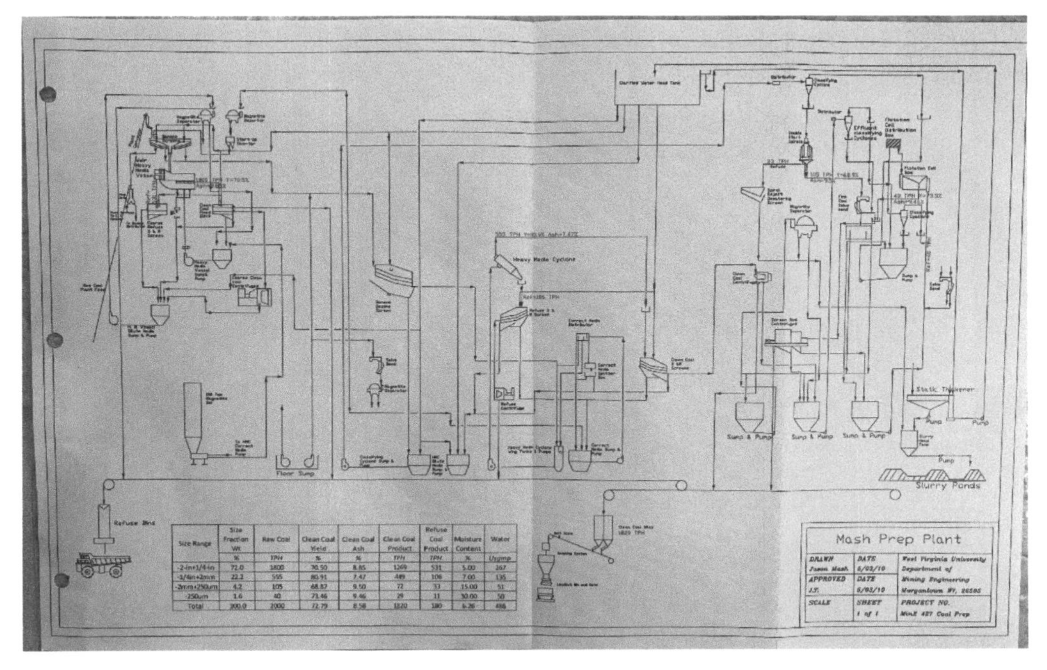

Figure 12.10 Example coal preparation plant flowsheet for a 6-million clean coal annual production mine

12.2 Solids and Water Balances

Given the sieve analysis of a raw coal feed in the following Table 12.1. A washability analysis shows a marketable coal can be produced if the separation specific gravity is at 1.4. The expected yield of clean coal is 70%. However, the ± 0.10 near gravity material is 30%. Draw

Table 12.1 Result of Sieve analysis

Size interval	Weight, %
+ 1-1/4 in	3
−1−1/4 +1/4 in	27
−1/4" + 28 mesh	20
−28 + 48 mesh	21
−48 + 100 mesh	10
−100 + 200 mesh	15
−200 + 325 mesh	3
−325 mesh	1

a flow sheet of coal preparation plant at 200 TPH feed showing estimated solid material flowrates, including solid-liquid separation circuits, but no other ancillary circuits. There is adequate area available for refuse disposal.

Figures 12.11 ~ 12.13 show three examples flowsheets and solid material flowrates.

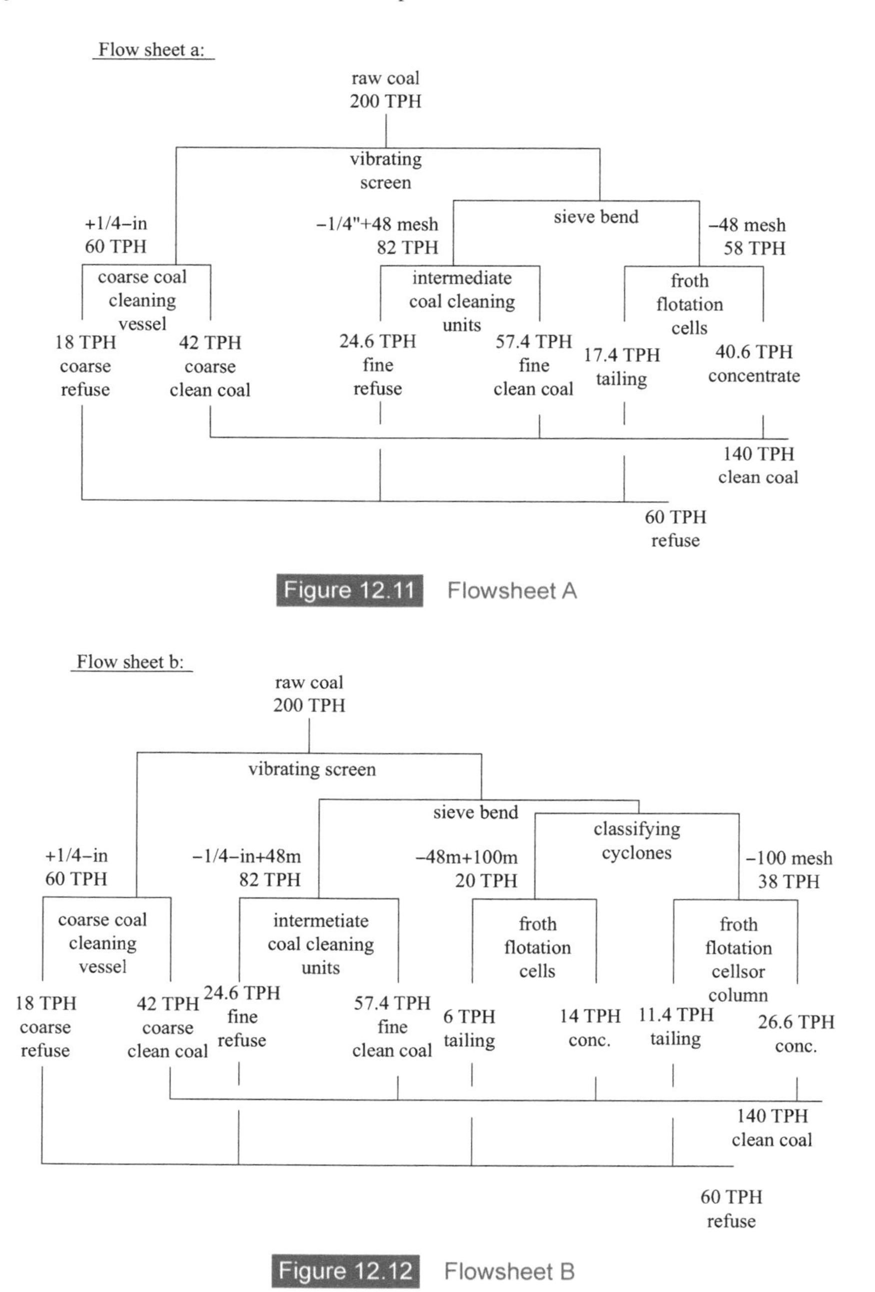

Figure 12.11 Flowsheet A

Figure 12.12 Flowsheet B

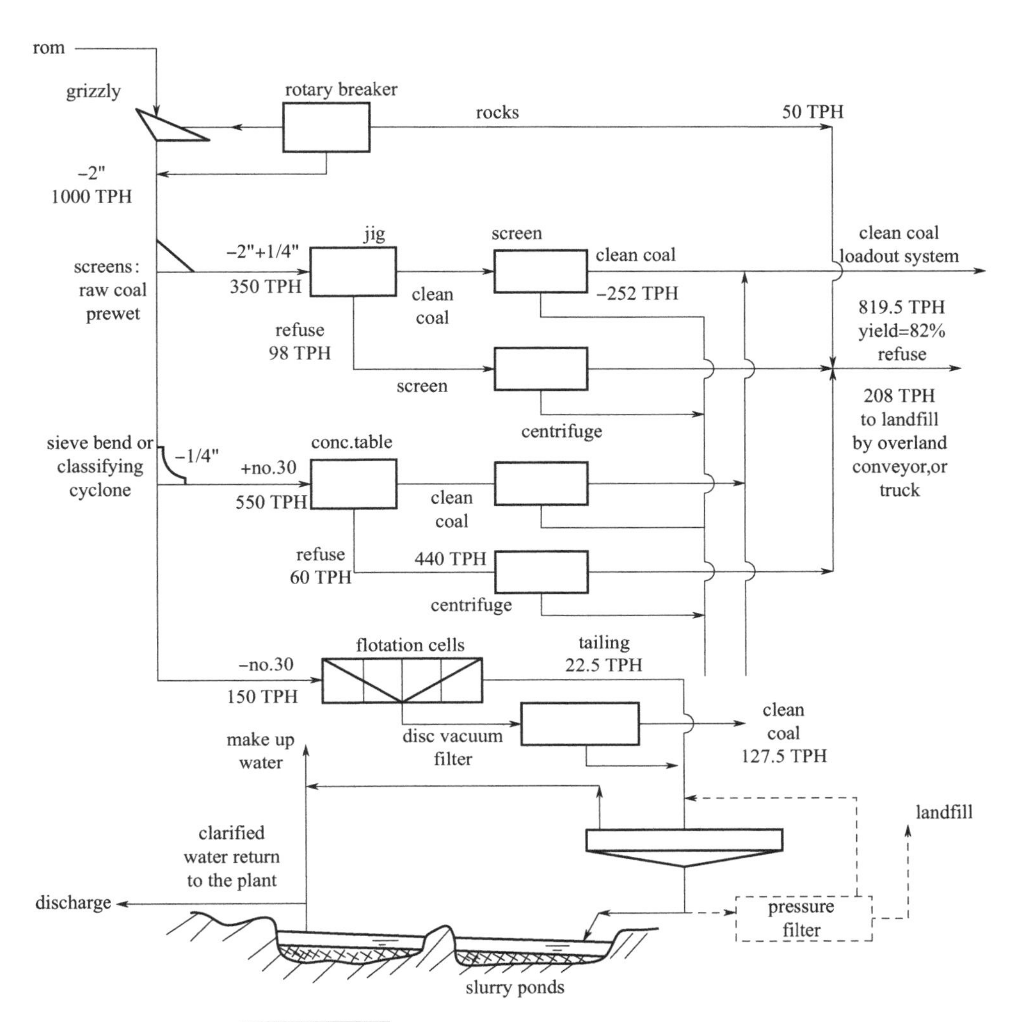

Figure 12.13 Coal preparation plant flowsheet C

COARSE REFUSE AND TAILINGS MANAGEMENT AND ENVIRONMENT

Tailings are the fine muddy or sandy mine wastes left behind after the valuable minerals have been extracted from the ores. Coal refuse is the material left over from coal mining, usually as tailings or spoil piles. The design, construction, and operation of tailings dams is a major consideration for most new mining projects, as well as for many existing mines. For a mine, the disposal of tailings added to total production costs, so it is essential to make its disposal cost-effective, in addition to satisfy the geological, technical, and environmental requirements.

13.1 Tailing Dams

13.1.1 Methods of Tailings Dam Construction

Three tailings dam construction methods, namely upstream, downstream, and centerline methods, are illustrated in Figures 13.1~13.3.

In the so-called **upstream method** of tailing dam construction (Figure 13.1), a small starter dam (or dyke) is placed at the extreme downstream point and the dam wall is progressively raised on the upstream side. The upstream method is named because the centerline of the dam moves upstream into the pond.

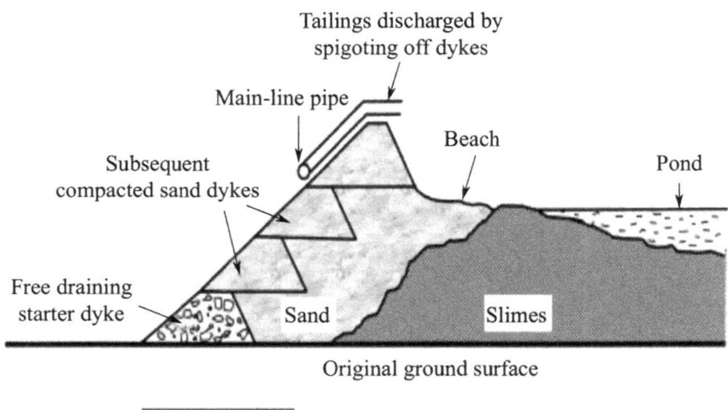

Figure 13.1 Upstream tailings dam

The **downstream method** has evolved as a result of efforts to devise methods for constructing larger and safer tailings dams. This method produces safer dams both in terms of static and seismic loadings. It is essentially the reverse of the upstream method in that as the dam wall is raised, the centerline shifts downstream, and the dam remains founded on coarse tailings (Figure 13.2).

The **centerline method** (Figure 13.3) is a variation of that used to construct the downstream dam and the crest remains in the same horizontal position as the dam wall is raised. It has the advantage of requiring smaller volumes of sand-fill to raise the crest to any given height. The dam can thus be raised more quickly and there is less trouble keeping it ahead of the tailings pond during the early stages of construction.

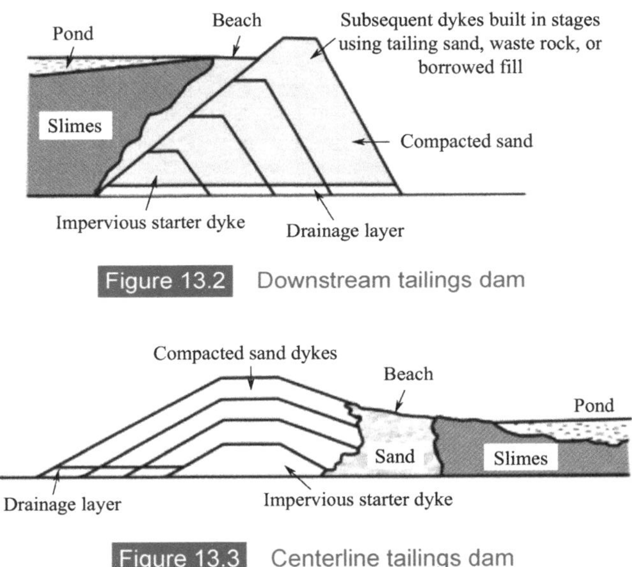

Figure 13.2 Downstream tailings dam

Figure 13.3 Centerline tailings dam

The upstream dams, built up on top of the tailings themselves, are the most common (and often cheapest) type of dams, and have significant risks of failure, especially in seismic and wet climate areas. The construction of new upstream tailings dams has already been banned in Brazil, Chile, Peru, and Ecuador. The centerline and downstream dams, built on crushed rock or fill, are generally less vulnerable to failures.

The site of tailings dams is usually planned close to the mine to save waste transportation cost. It is also cost-effective to build the dams across river valleys, if a suitable valley can be found close to the mine. The valley sides can serve as dam walls as shown in Figures 13.4 and 13.5.

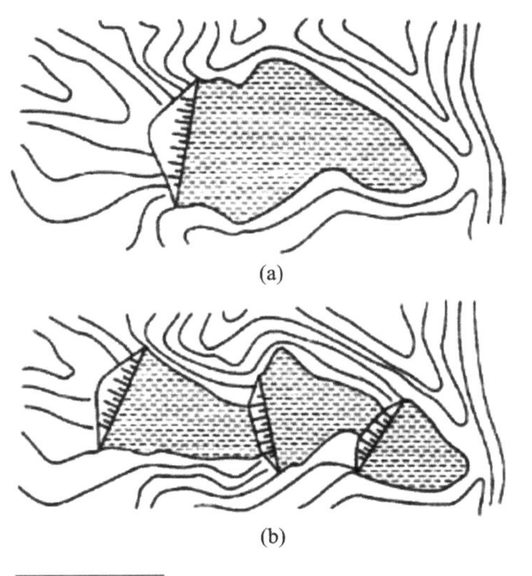

Figure 13.4 Cross valley impoundment single (a) and multiple (b)

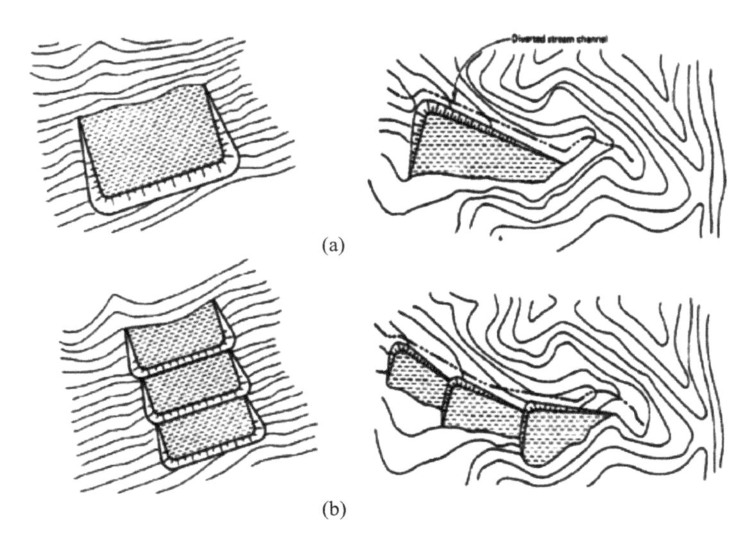

Figure 13.5 Side-hill and valley-bottom impoundment: single (a) and multiple (b)

13.1.2 Construction of Tailing Dam Wall Utilizing Cyclone Underflows

Hydro-cycloning of plant tailings is an often-used tool to produce relatively coarse free draining sand (cyclone underflow) for tailing dam construction. If separated out, the coarser refuse can serve as tailing dam construction material. It is generally more cost- and time-effective than truck transport of waste rock. More importantly the volume of tailings converted to dam construction material reduces simultaneously the tailings storage volume.

Figures 13.6 and 13.7 illustrate the construction of an upstream tailings dam using cyclone underflows. The wall is built from the coarse fraction of the tailings separated out by cyclone underflows, while the fines are directed into the pond.

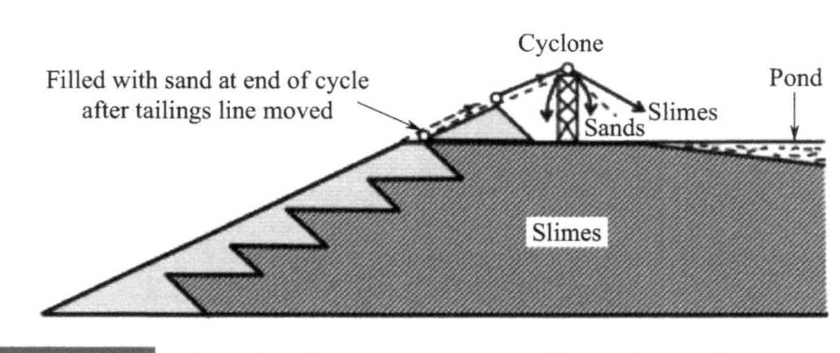

Figure 13.6 Construction of upstream tailings dam using cyclones

Figure 13.7 Construction of tailings dam wall using cyclone underflow

13.2 Contaminants from Reagents Used in Processed Plants

Several of the fluids and reagents used in coal processing plants either inherent or generated during coal flotation processing are toxic. Table 13.1 is a summary of the toxicity for reagents typically used in coal flotation.

Table 13.1 Toxicity of reagents used in coal flotation

Basic composition	Organism	Toxicity range	Toxicity
Polypropylene glycol	P. promelas	1000 to 10000 ppm v/v	Relatively nontoxic
C_4 to C_7 alcohols and hydrocarbon oil	P. promelas	32~30 ppm v/v	Moderate
Straight chain higher alcohols	P. promelas	10~100 ppm v/v	Moderate
Propylene glycol methyl eithers	P. promelas	> 1000 ppm v/v	Relatively nontoxic
MIBC	P. promelas	100 to 1000 ppm v/v	Relatively nontoxic
Xylenols, phenol cresols	Goldfish, chinok, silver and pink salmon	1 ppm	Killed in 6-48 hrs
Terpene hydrocarbons	Bluegill	3~7 ppm	Killed in 3 days
Ketones and alcohols	Fingerlings	46~49 ppm	Median lethal concentration

13.3 Water Balance in Tailings Impoundment

13.3.1 Water Gain and Loss in Impoundment

A generalized consideration of water gain and loss in a tailings impoundment includes the following elements: Slurry water, Precipitation, Evaporation, Surface run-off, Resurgence to subsoil and groundwater, Resurgence below dam, Seepage, Decanting and water reclamation.

With the exception of precipitation and evaporation, the rate and volume of water can be controlled to a large extent. If surface run-off to the dam is substantial, then interception ditches should be installed. Recycling of water from the decant is becoming more important due to pressure from governments and environmentalists. The difference between the total volume of water entering the tailings pond and the volume of water reclaimed plus evaporation losses must be stored with the tailings in the dam. If that difference exceeds the volume of the voids in the stored tailings, there is a surplus of free water that can build up to tremendous quantities over the life of a mine. The main disadvantage of water reclamation is the recirculation of pollutants to the process plant, which can interfere with processes such as flotation. The main effects of pollution are:

① the effluent pH (such as acid mine drainage) may cause ecological change;

② the dissolved heavy metals, such as copper and lead, can be lethal to fish-life if allowed to enter local water courses;

③ the reagents, which are usually present in very small quantities, nevertheless may be harmful; and

④ the suspended solids which should be minimal if the tailings have spent long residence time in the dam, thus allowing the solids to settle and produce a clear decanting water.

A number of wastewater treatment techniques are available, such as physical adsorption methods using coal or bentonite clay or mineral slimes, limestone or lime treatment, biological oxidation of organics, removal of ionic species by ion exchange resins, and reverse osmosis. The mineralogical nature of the tailings often provides natural pollution control. For instance, the presence of rich alkaline gangue minerals such as limestone can render metals less soluble and neutralize oxidation products. Such coal mines thus present less problems than the tailings with rich in pyritic gangue minerals.

13.3.2 Water Reclamation System for Closed-Water Circuit

A closed water circuit for coal preparation plant and tailings impoundment is important from the environmental control of contaminants and conservation of water point of view. As much water as possible must be reclaimed from the tailings impoundment for re-use in the coal preparation plant and the volume of fresh make-up water used must be kept to minimum. The water balance including total volume of water entering the tailings impoundment and the volume of water reclaimed plus evaporation losses must be calculated. The difference must be stored in the tailings impoundment. They might become a surplus of free water that can build up to tremendous quantities over the life of the mine. A typical tailings impoundment water recirculation system is given in Figure 13.8, while a typical dam-reclaim system is shown in Figure 13.9.

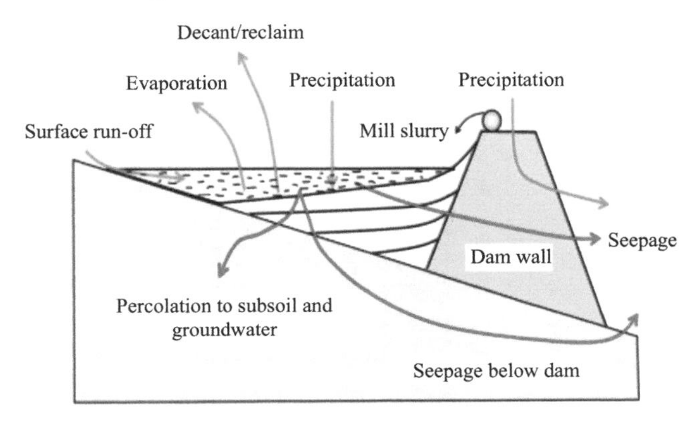

Figure 13.8 Water gain and loss in a typical tailings dam

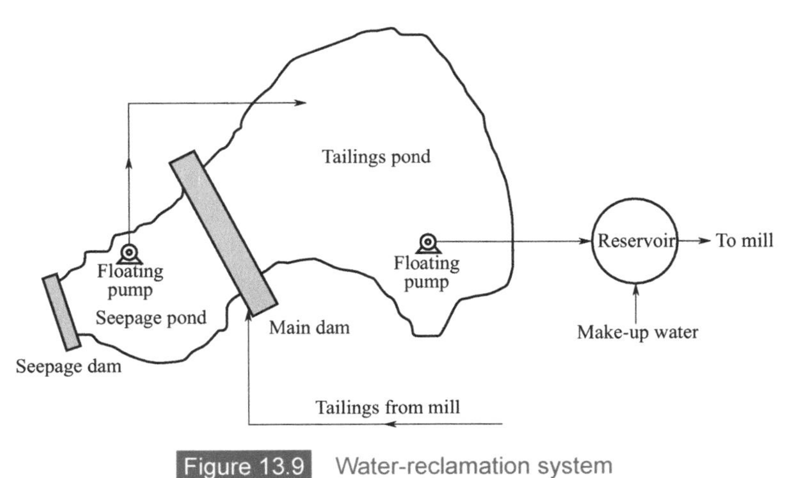

Figure 13.9 Water-reclamation system

13.3.3 Design of a Settling Basin – An Example

Example problem: A coal preparation plant discharges 20,000 gallons per hour of tailings containing 5 percent solids (the solid concentration in the effluent may vary from 5 to 20 percent) with the plant operating 6 days per week. It is desired to determine the size of the settling pond required.

Assumptions

① 45 ft^3/ton of fines is used to estimate the storage volume of fines to be deposited.

② The volume available to store the fines should be sufficient for a six-month period as a minimum.

③ There available at least 8 hours settling time in the sedimentation zone.

④ The most desirable shape of the impoundment has been suggested to be 3:1 length to width ratio of a rectangle shape.

⑤ It can accommodate a maximum 3 ft water depth.

⑥ Unit conversion factors: 7.5 gal/ft^3 and 2,000 lb/ton.

Solution:

① The weight of fines in the waste slurry can be calculated as follows:

Fines in lb per day (assuming that the solid concentration is the average of 5% and 20% volume by volume):

= flowrate (gal/hr) × 24 hr/day × solid fine concentration / 7.5 gal/ft^3 /45 ft^3/ton × 2,000 lb/ton

=20,000 × 24 × (5% + 20%)/2 / 7.5 / 45 × 2,000

=355,556 lb/day

Fines in ton per day (tpd)

=355,556 / 2,000 = 178 ton/day

Fines volume storage per day:

=178 ton/day × 45 ft^3/ton = 8,000 ft^3/day

8,000 ft^3/day / $(3\ yd/ft)^3$ = 296 yd^3 storage required per day

Fines volume storage for 6 months

=296 yd^3/day ×6 days/week × 4.25 weeks per month × 6 months

= 45,333 yd^3 storage volume required

Assume solid fines depth of 12 ft or 4 yd

Storage surface area = 45333 yd^3 / 4 yd = 11,333 yd^2

Using approximately 3∶1 length to width ratio:

Assume width=y, then length = 3y

3y × y = 11,333 yd^2

Therefore width = 61.5 yd (use 62 yd), length = 184.4 yd (use 184 yd)

Designed volume for storage 6-month solid fines:

= 4 yd × 184 × 62 = 45,632 yd^3

= 12 ft deep × 552 ft long × 186 ft width = 1,232,064 ft^3

② Settling area required for water

20,000 gal/hr × 8 hr settling time

= 160,000 gal storage required

160,000 gal / 7.5 gal per ft^3 = 21,333 ft^3

Water depth to be 3 ft maximum, thus

Minimum surface area required

=21,333 ft^3 / 3 ft = 7,111 ft^2

Assume shape of settling pond roughly 3∶1 ratio (pond width = z, length = 3z):

$3z^2 = 7,111\ ft^2$

z = 48.7 ft (use 50 ft)

Therefore, the pond surface area = 150 ft long × 50 ft wide = <u>7,500 ft^2 surface area</u>

HOMEWORK, PROJECTS, LABORATORY EXPERIMENTS, AND EXAMS

14.1 Homework

Homework #1 Coal Resources and Characterization

1. ① Why is coal preparation necessary for run-of-mine coal?

② What are the commonly known techniques used in coal beneficiation?

2. What can coal be used for (coal utilization)?

3. What are the major coal resource fields and regions in U.S.A?

4. What are the major coal seams in northern and southern coal fields of West Virginia? List 5 major seams in each coal field.

5. What transportation methods or modes can be used to deliver

① run-of mine coal from underground or surface mining site to the crushing plant of coal preparation plant?

② raw coal or clean coal products directly delivered to the end users.

③ What are the names of major waterways (river systems) used to transport coal to markets in U.S.?

6. What is the Parr formula for defining the mineral matters in coal?

① What are the bases, and the Parr formula, used for coal classification by rank in U.S.?

② What is the agglomerating property of coals in class and groups of coal in U.S.? What is the significance of this property?

③ Name the class and rank of coals that have this property.

7. Name the ranks of coals in U.S. coal fields.

8. ① What are the proximate analysis and ultimate analysis involved in chemical analysis of coal?

② How is the total sulfur determined?

③ Describe in details the forms of sulfur in coal, and the forms of organic sulfur in coal.

9. How are the following coal properties determined: a. total moisture, b. volatile matters, c. fixed carbon, and d. oxygen.

10. ① Define the Hardgrove Grindability Index (HGI) and describe how to determine the HGI of a given coal per ASTM Standard.

② Name two commonly mined coal seams, two each in WV, OH, and PA for high HGI (about 100), medium HGI (about 60), and low HGI (about 45).

11. Given a coal with 2.75% sulfur, 0.488 ppm Hg, and 13,700 Btu/lb, determine,

① the sulfur dioxide emission rate in SO_2 lbs/million Btu.

② the mercury emission rate in Hg lbs/trillion Btu. From the result of Hg content, and find out which region of coal might contain the same amount of mercury.

Homework #2 Environmental Impact of Coal Utilization, and Crushing and Comminution

1. ① What are the major environmental regulations that govern the emission of air quality standards due to combustion of fossil fuel such as coal in power plants and industrial boilers?

② What are the major restrictions or limitations of those legislative regulations addressed the use of fossil fuel?

2. Name one U.S. Federal Laws and one Amendment that are intended to reduce air pollution and protect the air quality. What are the major pollutants stated in the law and amendment require them to be controlled and regulated.

3. What is the major law or global agreement that has been agreed upon for carbon management. What are the major approaches that can be taken to achieve the goals.

4. ① What are the purposes for reducing the size of mineral particles using the comminution process?

② What are the major principles for the comminution process?

5. ① List the names of unit operations commonly used for crushing coal in coal preparation plant?

② What is the basis used to select the size reduction unit operations for different types of coal?

6. List the names of major unit operations used for grinding and pulverizing coal?

7. What is the particle size in micron size unit (μm) and Tyler mesh unit for the following US sieve series:

USA Sieve	Micron Size, μm	Tyler Unit, Mesh
USA sieve series no. 16		
USA sieve series no. 18		
USA sieve series no. 30		
USA sieve series no. 50		
USA sieve series no. 60		
USA sieve series no. 100		
USA sieve series no. 200		
USA sieve series no. 325		
USA sieve series no. 400		

8. Name the unit operations and corresponding size ranges (can be processed) for sizing, classifying, and desliming coal in coal preparation plant.

9. To estimate the energy requirement for size reduction, there are three criteria and combined criteria can be used. Briefly describe the basis of those criteria.

10. What are the purposes of particle sizing?

Homework #3 Material Balances in Crushing Plant and Grinding Circuit

1. Use 1 ton medium as base to calculate the liquid requirement for preparation of a dense medium with a specific gravity of 1.55. Determine the ratio of solid to liquid requirement by weight. Assume magnetite has a specific gravity of 4.95.

2. Material balance of a grinding-classifying closed circuit;

① Determine the material balance in terms of TPH, specific gravity, and gpm for solids, liquid, and slurry in a grinding-classifying closed circuit shown in Figure 5.2. The condition of the circuit is similar to the example given in the lecture and textbook, except the specific gravity of the solid is 3.45, and feed to the circuit is 800 TPH dry with surface moisture of 9%.

There are major nodes needs to be considered for the closed circuit to balance the solids and water. Take the materials balances around these nodes:

a. Overall (whole circuit) closed grinding-hydrocyclone circuit.

b. Classifying cyclone.

c. Joint, junction, or blending point of the underflow from classifying cyclone and plant feed to form a new feed for the ball mill.

② Show the detailed calculations of material balances including the intermediate steps. You are required to use an Excel spreadsheet for the calculations and attach copies of the Excel sheets of the detailed calculations.

③ Show the summary of results in the "material balance tables", and type in legends for the grinding-classifying cyclone close circuit given in Figure 5.2.

④ Calculate the make-up water for the underflow of the cyclone to meet the solid concentration required for the grinding mill, and for the sump to meet the solid concentration requirement for the feed to the classified cyclone. Also calculate the overall amount of make-up water requirement to the circuit.

Homework #4 Washability Data and Prediction of Raw Coal Cleaning Results

1. Explain the advantage of using hinder-settling based separation compared to free-

settling based separation, using the concentration criteria for low and high Reynolds number flow regions.

2. What are the three major sub-motions of the settling velocity of a particle in fluid?

3. What are the critical factors for enhancing the performance in solid-solid separation in increasing the differences between initial acceleration and settling velocity of lighter and heavier particles?

4. Define the sphericity of a particle and determine the sphericity of a given particle that has the shape of width = d, height = 0.5 d, and depth = 2d.

5. Using the settling velocity Equation (7-12a) in p. 113 of the text, calculate the settling velocities for the following flowing particles in Excel Worksheet. Plot all settling velocity curves are similar to the example in the chart given in the textbook. Discuss and conclude the results.

① 0.3 mm spherical coal particle.

② 0.3 mm spherical quartz particle.

③ 0.5 mm spherical coal particle.

④ 0.3 mm cubic coal particle.

⑤ 0.389 mm spherical coal particle.

6. To prepare a dense-medium with specific gravity 1.65 for dense medium cyclone separation, and determine the weight ratio, and volumetric ratio of magnetite to water. The magnetite has a specific gravity of 5.2.

7. (a) Name the commonly use two concentrators for beneficiating raw coal to clean coal product and refuse (impurities) in each coal cleaning circuit in coal preparation plant.

(b) What are the corresponding raw coal size ranges for each unit operation (concentrator) selected in each coal cleaning circuit.

8. Name two most commonly use concentrators used for dry raw coal cleaning operations.

9. What methods are used to recover dense medium such as magnetite after concentration of raw coal in coal preparation plant.

Homework #5 Design Project

Determine the fine coal flotation yield for ultrafine coal cleaning circuit and calculate the material balance in TPH for four coal cleaning circuits and whole process plant and surface facilities.

1. Complete lab report #4 for "Flotation of Fine Coal."

2. Use the overall yield and ash from flotation of fine coal data for clean coal yield% and

ash% of floatation results for coal preparation plant flowsheet material balances. However, the overall yield will be discounted by a factor of 0.82. This means

Actual flotation yield% = (flotation batch test clean coal yield%) × 0.87

3. Construct a table similar to the one shown below and fill up all the necessary data. Then, copy and paste the completed table into your AutoCAD coal preparation plant flowsheet (such as below the area of the crushing plant). The moisture contents for clean coal product from each cleaning circuit are assumed in the table.

4. You must describe the purposes of the detailed calculation work for each assignment for the final coal preparation plant design project report to receive the full credit (do not just attached calculations).

5. Save all of your lab reports applicable to the coal preparation plant design project, and assignments for Coal Preparation Plant Flowsheet Design and Material Handling Facilities files for your final coal preparation plant flowsheet design project report. Save all of your work in a CD. Scan the graphs as e-files and also included in CD, such as Rosin-Rammler's graph, washability curves, etc. This is part of your Capstone Mine design project report. You must have descriptions and organize for the final plant design final report, including cover page, table of contents, introduction, etc. for full credit. Add figures and tables with sequential numbers.

6. Refine your coal preparation plant flowsheet. Print out the coal preparation flowsheet in 11"×17" size sheet.

Size, %	Feed to each cleaning circuit, TPH	Clean coal yield, %	Ash, %	Moisture, %	Refuse yield, %
				5	
				12	
				25	
				30	
Total					

Homework #6 Flotation of Fine Coal and Oxidized Coal

1. List the commonly used frothers and collectors for flotation of fine coal in coal preparation plants. What are the functions of those reagents in the flotation of fine coal.

2. List the commonly used froth flotation unit operations (devices) used in coal preparation plants.

3. Describe the surface properties of coal particles and that of refuse particles such as particles rich in silica, that are major differences used in separation of fine coal from mineral matters in a flotation process.

4. Read out the data from the textbook, and construct the tables to include the carbon contents of coal versus contact angles of coal solids, and the ash% of coal versus contact angles of coal solids.

5. (a) For the solid-liquid-gas interface system, derive the Young's Equation to describe the relationship between surface tension and contact angle.

(b) Derive the Dupree's Equation of free energy for the attachment process.

(c) Derive the change of Gibb's Energy for being the function of surface tension of liquid-solid interphase, and contact angle between solid surface and air bubble.

6. In order for a solid particle to attach to the surface of an air bubble, what conditions or requirements must be met?

7. Rank the floatability of coals in the United States is based on the rank of coal.

8. (a) List the types of coals that might be suspected to be oxidized coal under certain conditions.

(b) Explain the possible reasons for coal to be oxidized.

9. What kinds of chemicals or functional groups are predominantly covered on the surface of the oxidized and non-oxidized coals?

10. List the adverse effects of coal oxidation on coal preparation.

11. What are the major mechanisms for the rate of fine coal flotation and the expression of the flotation rate?

Homework #7 Dewatering and Thickening and Online Quality Monitoring

1. Name the commonly used filtration unit operations for dewatering of fine clean coal and ultrafine clean coal products in coal preparation plants.

2. Name the dewatering unit operations used for fine refuse and ultrafine tailings in coal preparation plants.

3. Name two unit operations that can be used for dewatering of the fine coal tailings for the purpose of volume reduction of tailings to be stored in the impoundment.

4. What unit operation has been used for thermal drying of coal?

5. To concentrate the effluent (black water) from a high frequency banana screen and a screen bowl centrifuge, what unit operation is most commonly used in coal preparation plant?

6. What unit operation is commonly used to clarify the black water for reuse of the process water in the coal preparation plant?

7. What are the most commonly used flocculants (non ionic) and reagents (cationic) to be used as dewatering aids in coal preparation plants?

8. What is the major mechanism of flocculant function?

9. What is the application of on-line coal quality monitoring systems?

14.2 Laboratory Experiments

Laboratory #1 - Preparation of "representative" Coal Sample and Chemical Analysis

1. Background

Coal analyses include ultimate and proximate analyses and determination of forms of sulfur, and calorific value.

Proximate analysis test has proved very useful because it supplies a single and extensively used characterization of coal. This test includes the analysis of total moisture, volatile matter and ash contents in coal sample and calculation of fixed carbon.

Sample preparation consists of systematically reducing the size of the coal sample for size or chemical analysis without altering its properties.

2. Objectives

The purpose of this experiment is to learn how to properly prepare the samples of coal for chemical and particle size analyses and additional experiments. To meet these goals, students will:

(1) Perform sample reduction using alternate shovel, cone and quarter techniques.

(2) Learn to use laboratory-scale crushers, grinders and mill.

(3) Learn to prepare the coal samples for chemical analysis.

- Compare the analysis of two independent sub-samples taken from a single gross sample.
- Prepare coal samples for the following tests and analysis:

① Proximate and ultimate analyses and determination of heating value.

② Particle size distribution and analysis.

③ Float and sink test and analysis.

④ Performance evaluation for wet coal cleaning unit or circuit (assignment for plant design project).

⑤ Froth flotation of fine coal.

⑥ Sedimentation rate and thickening.

⑦ Save (reserve) for additional tests.

3. Equipment

3.1 Laboratory Equipment

- No. 4 scoop shovel.

- Hammers.
- Jaw crusher.
- Double roll crusher.
- Grinding mill.
- Holes mill (-U.S. No.60, 250 um).
- Coarse and fine rifflers (sample dividers).
- Various size of coal sample plastic bags.
- Furnace.
- Draft oven.
- Total sulfur analyzer.
- Top loading balance and analytical balances.

3.2 Personal Equipment and Safety Training

- Lab coat.
- Safety shoes.
- Safety glasses.
- Hair cap.
- Respirator or dust mask.
- Leather gloves and disposable gloves.
- Training handbook for using Crusher Lab.

3.3 Clean-up equipment and lab: Always clean up the labs after experiment

- Liquid soap and paper towel.
- Floor brush or broom and vacuum cleaner.

4. Experimental Procedure

The class will work as a team and start with an air-dried gross sample of coal. Each student will prepare two samples for chemical analysis by the following six steps listed in Section 4.2 below.

4.1 Materials

Waynesburg seam coal (a bituminous coal), Monongalia County, WV

4.2 Procedures

4.2.1 Coal sample volume reduction

Long pile and alternative shovel method

- As the coal sample size is larger than 2-in size, use the hammer to break it to about 2-in size.
- Spread the gross sample into a line about 3-ft wide by 6-in high.
- Use No.4 scoop shovels to take alternate shovel widths of coal.
- Place the alternate scoops of coal into a cone shape on the floor. This is cone A.
- Shovel up the reject gross sample and form it into cone B on another section of the floor.

Cone and Quarter Technique

- Use the scoop shovel to flatten out cone A into a pancake shape about one foot high.
- Use the shovel to divide each pancake into 4 equally sized quarters.
- Designate the quarters of each cone as: cone A: 1-A, 2-A, 3-A, 4-A.
- Store 1-A into a container for particle reduction later.
- Store 1-B into a sieve for size distribution analysis later.
- Store other samples in plastic bags for later use.

4.2.2 Stage size reduction

Crushing and Screening

- Sieve sample 1-A through the 1-in screen to obtain a plus 1-in fraction and a minus 1-in size fraction.
- Sieve 1-A's minus 25.4mm (-1-in) fraction through a 4 mesh screen to get a -1-in +4 mesh size fraction, and a minus 4.7mm (4 mesh Tyler) size fraction.
- Adjust the throat of the jaw crusher to 1 inch.
- Crush 1-A's plus 1-in fraction in jaw crusher. Combine the jaw crusher discharge with the −25.4mm+4.7mm (−1-in+4mesh) size fraction.
- Adjust the double-roll crusher to about 6.35 mm (1/4 in).
- Crush 1-A's −25.4mm+4.7mm(−1-in+4mesh) size fraction in the double-roll crusher. Combine the double roll discharge with the minus 4-mesh fraction.

Coarse Riffling and Grinding

- Riffle sample 1-A with the coarse riffler, eliminating one half of the remaining sample with each pass until two splits about 3~5 lbs or smaller sample are left.
- Retain one 3~5 lbs sample split. Store the other split and the remaining rejected and labelled it "1-A reject".
- Use the grinding mill to further reduce the particle size of double roll crusher discharge. Clean the mill thoroughly.

Pulverizing and fine riffling

- Pulverize sample 1-A in the Holmes mill (the screen plate is for 250 μm, No.60 or 60 mesh Tyler).
- Riffle sample 1-A with the fine riffler, eliminating one half of the remaining sample until two 50-gram splits remains.
- Place one of the 50-gram splits in a sample jar and label it "1-A".
- Discard the reject.
- Clean the mill completely.

4.2.3 Analysis of chemical components - proximate analysis

Surface moisture

- Preheat the oven to about 150℃.
- Weigh 1 gram of the coal sample A1 and A2 separately on a weighed grass plate.
- Put the samples in the oven for 1 hour.
- Remove the samples from the oven and weigh them after they are totally cooled down.

Ash

- Preheat the oven to about 750℃.
- Weigh 1 gram of the coal sample A1 and A2 separately in a weighed dish.
- Put the samples in the muffle oven for 4 hours.
- Remove the samples from the oven and weigh them after they are totally cooled down.

Volatile matters

- Preheat the oven to about 950℃.
- Weigh 1 gram of the coal sample A1 and A2 separately in a weighed platinum crucible.
- Put the samples in the furnace chamber for 7 minutes.
- Remove the samples from the oven and weigh after they are totally cooled down.

Sulfur

- Preheat the oven to about 1000℃.
- Weigh 0.3 grams of the coal samples A1 and A2 separately in a weighed boat.
- Put the samples in the sulfur meter for 3 minutes.
- Record the sulfur value directly from the sulfur meter.

4.2.4 Clean-up Laboratories

Clean all equipment, containers, table space, and floor space used in the lab, using brushes and wet towels.

Laboratory #2 - Size Distribution and Analysis

1. Background

Accurate size consisted data is essential in designing the flowsheet for a coal preparation plant. Plant designs seek to maximize the product yield while meeting a particular customer specification.

All plant equipment is sized from the projected coal throughput of each of the three typical circuits (coarse, fine, and ultrafine)

The most widely used measure of particle size for coals is sieve size.

2. Objectives

The purpose of this experiment is to learn how to perform and interpret the results of a size analysis of a coal sample, and describe the size distribution in mathematical expression. To meet these goals, students will:

(1) Perform size analyses of coal sample obtained from a single gross sample of coal using the nested selected sieves on Ro-tap to reduce its volume.

(2) Tabulate the sieving test results in Excel Worksheet. Plot the Rosin-Rammler (R-R) curve on the Rosin-Rammler-Bennet graph paper to determine the parameters of the R-R equation.

(3) Linearize the Rosin-Rammler equation. Then use the Trend-line in the worksheet to obtain a least square fitted linear curve of the R-R equation and determine the parameters of the R-R equation.

(4) Study the characteristics of the R-R distribution and compare the size distribution results obtained between the graphic method and the analytical method.

3. Equipment

(1) Laboratory Equipment

① Test Sieves (1 in, ¼ in, U.S. Nos 30, 50, 100, and 200) and pan.

② Ro-Tap machine.

③ Top loading and analytical balances.

④ Shovels and sample scoops.

⑤ Sample plastic bags and marker.

(2) Personal Equipment and Safety Training

① Lab coat or old clothes.

② Safety equipment (safety shoes and glasses, dust mask, gloves).

4. Experimental Procedure

The class will work in groups to perform size analysis of coal samples that are obtained in Lab #1.

(1) Machine sieve a manageable batch of sample 2A on Ro-tap for 20 minutes, using a nest of sieves as follows: bottom pan, 150 μm (100 mesh Tyler), 250 μm (60 mesh), 600 μm (28 mesh), 6.35 mm(1/4-in), 25.4 mm(1-in) etc., and lid.

(2) After each batch has been sieved, carefully separate the sized fractions into pre-weighted containers or plastic sample bags.

(3) When the entire sample has been sieved, weigh each loaded container to the nearest 0.1 g. You should record the net weights, with the tare weight subtracted out.

(4) Store each size fraction in a clearly labeled plastic sample bags.

(5) Clean-up laboratory and equipment.

(6) Thoroughly clean all equipment, containers, table space, and floor space using brushes, wet towels, and vacuum cleaner.

5. Data Analysis and Report

Use a spreadsheet for data reduction and analysis, and plotting graphs. Type an experiment report using a word processor including background, objectives, experiment procedures, results, discussion, and references.

For size data analysis, see examples shown in the textbook and illustrated in the class.

① Calculate r(x) and R(x) values for all of the size fractions of samples.

② Make a table to include sizes in US sieve, Tyler mesh, and SI units, direct weight in grams, individual weight in fraction, r(x), and cumulative weight in fraction and percent, R(x).

③ Make another table to include size x, in mm, R(x) in %, ln (ln(100/R(x))), if R(x) % is used.

(1) Graphic Method

① In Rosin-Rammler-Bennet graph paper, plot x vs R(x) for the sample.

② Draw a straight line for the size data using the best-fit concept. The straight line can be one or two lines. Two straight lines represent a bi-model size distribution, otherwise it is a uni-model.

③ Determine two parameters, the slope n and the absolute size constant x_o.

④ Determine the Rosin-Rammler equation .

(2) Analytical Method

① Use the built in analytical tool in spreadsheet for linear regression analysis to fit the Rosin-Rammler equation to the observed particle size data.

② Calculate each term in the linearized Rosin-Rammler equation,

$\ln(\ln(100/R(x))) = -n\ln x_o + n\ln x$

$Y = b + mX$

where Y is ln(ln(100/R(x))), X is lnx, and m = n is slope, $b = -n\ln x_o$ is intercept.

③ In spreadsheets such as Excel, plot X vs Y in a scatter figure.

④ Find the sum of square of deviation of data points.

⑤ If bi-model, fit the Rosin-Rammler equation into two separate straight lines equations.

⑥ Compare the Rosin-Rammler equation developed from these two methods.

Laboratory #3 - Washability Analysis of Coal

1. Background

In coal washability, the objective is to establish the set of maximum possible separation performance criteria for a given coal feed. A number of these curves are drawn to illustrate different conditions or variables, usually on the same axes, thus presenting the information on one sheet of paper. Washability curves are essential when designing a new coal or mineral preparation plants. There are four main types of washability curves: characteristic ash curve, cumulative float curve, cumulative sink curve, and densimetric or specific gravity curve. The different curves are based on the relation between ash or ore grade, mineral density, and particle size distribution.

The washability of coal or minerals is expressed by a curve or graph showing the results of a series of float-and-sink tests.

2. Objectives

The purpose of this experiment is to learn how to separate the coal sample into different specific gravity ranges; obtain the ash content of each density ranges and do the washability analysis with the assistance of the washability graph.

3. Equipment

(1) Laboratory Equipment

① 1/4-in + U.S. Series No. 16 (-1/4-in +1 mm) coal sample.

② Plastic sample bags.

③ Top loading laboratory balance.

④ 200 ml beakers or float-sink flasks with stoppers.

⑤ Hydrometer, 2000 ml graduated cylinder, 2-quart or 4-quarts Pyrex glass trays.

⑥ Constant Lithium Metatungustate Liquids (LMT).

⑦ Prepare LMT liquids of 1.30, 1.40, 1.50, 1.60, 1.70 and 1.80 specific gravity.

⑧ Stainless steel wire mesh strainers.

⑨ Vacuum filtering systems including, Buchner funnels, No. 25 filter paper, Tygon tubing, 200 ml and 3000 ml Erlenmeyer flasks, T connector, and vacuum line (building).

(2) Chemical Analysis of Coal

① Laboratory analytical hood.

② Analytical balances.

③ Holmes mill.

④ Programmable furnace, crucible, weighing papers, desiccators.

⑤ Total sulfur analyzer, Parr calorific meter.

(3) Personal Equipment and Safety Training

① Lab coat.

② Safety shoes.

③ Safety glasses.

④ Hair cap.

⑤ Respirator or dust mask.

⑥ Leather gloves and disposable gloves.

⑦ Training handbook for using Crusher Lab

4. Experimental Procedure

4.1 Materials

Waynesburg seam coal (a bituminous coal), Monongalia County, WV

4.2 Procedures

① Weigh 200-30 grams of −1/4-in. +U.S. No. 16 (−1/4-in +1 mm) fraction of coal sample, using the top loading balance.

② Put the coal sample to the dense medium with a specific gravity of 1.3; stir the coal to ensure it has been wetted. Leave the coal to sink and float for half an hour, use the steel strainer to collect the coal floating in the surface of the dense medium, and then filtrate both the float and sink coal.

③ Put the float into the furnace for one hour to eliminate the water content. Pour the sink coal into the higher dense medium of specific gravity 1.4. After 15 minutes, the float is gathered with a strainer and then filtrated both the sink and float separately.

④ Put the float into the furnace for one hour and put the sink into a higher dense medium with the specific gravity of 1.5. Repeat step two to step three in 0.1 specific gravity until it reaches 2.0. The sinks of one dense medium should be put into a higher dense medium for sink and float flotation. The specific gravity of each dense medium is 1.3, 1.4, 1.5, 1.6, 1.7, 1.8, and 2.0. Put the floats of each sink-float experiment and the sink of the last sink-float experiment into the furnace for one hour to eliminate the water content.

⑤ After finishing the sink-float experiments, the original coal sample is splitted into eight different coal samples; they are separated into the following density ranges: −1.3; 1.3-1.4; 1.4-1.5; 1.5-1.6; 1.6-1.7; 1.7-1.8; 1.8-2.0; +2.0.

⑥ Turn on the muffle furnace and place around 1g coal sample of each density range in the burning dishes. Record the weight of the furnace dish, coal sample and total weight accurately, and then put the burning dishes into the muffle furnace. When the temperature of the muffle furnace rises to 750℃ , open the door of the muffle furnace a little to allow the air to flow into the furnace for one hour. Finally close the muffle furnace door for three hours. Get the burning dishes out, weight each of them and calculate the ash weight of each density range. Record the ash content in percentage of each density range.

⑦ Clean all equipment, containers, tablespace, and floor space used in the lab, using brushes and wet towels.

5. Data Analysis and Report

Use spreadsheet software tools for data analysis and PSI-PLOT graphic software for plotting the washability curves, and work processing for report presentation. The report must include background, objectives, experimental procedures, results, discussion, and references.

(1) Using washability analysis method to calculate the direct values in weight percentage (dry basis) at each specific gravity interval. In addition, calculate the cumulative float and cumulative sink weight percentages, and Z-ordinates for ash content.

(2) Plot the following five washability curves on a single figure:

① weight (yield) specific gravity curve.

② cumulative float curves for ash and sulfur.

③ cumulative sink curves for ash and sulfur.

④ Element curves for ash and sulfur.

⑤ ±0.10 specific gravity distributions.

(3) The prospective customer has requested −1/4 in + No.30 US sieve coal that has a maximum ash content of 7.0% and a maximum sulfur content of 1.7%. How should this coal be cleaned to meet these specifications? What is the maximum theoretical yield obtainable?

(4) For the specific gravity used to clean the coal (determined from Step 3), determine the followings:

① the theoretical clean coal yield (cumulative float weight percent), ash and sulfur contents.

② theoretical ash and sulfur contents of the refuse.

③ the highest ash and sulfur content of any individual particle found in the float coal product.

④ the percentage of total raw coal feed which lies within ± 0.10 specific gravity unit of the specific gravity of separation.

Laboratory #4 - Froth Flotation

1. Background

Froth flotation is based on the difference of surface properties between coal and mineral matters to achieve the separation. The coal surface is mostly covered by non-polar groups which consist mainly of carbon and hydrogen. While the surface of mineral matter is covered by polar groups which mainly consist of carbon, oxygen, hydrogen and mineral matters. The surface of coal particles presents hydrophobicity, meaning it can be attached to an air bubble with the assistance of some chemical reagents such as a collector and a frother. The surface property of mineral matters is reported to be hydrophilic. So when the flotation begins, the coal particles will be attached to the air bubbles and be moved to the frother layer at the top, while the mineral matters will remain in the bottom of the flotation cell.

2. Objective

The purpose of the experiment is to study the basic principle of froth flotation. To achieve the goal, a bench-scale flotation experiment on ultrafine coal less than 1 mm is conducted to determine the recovery percent and ash percent of each time interval and at the cumulative time points. Finally, the recovery of combustible material of the clean coal at each cumulative time point is determined.

3. Equipment

(1) Laboratory Equipment

① Riffler or sample splitter.

② Top-loading and analytical balances.

③ 2000 ml graduated cylinder.

④ Laboratory agitator.

⑤ Automatic Denver laboratory D-12 flotation machine.

⑥ 2-liter stirred tank type flotation cell.

⑦ 1 ml volumetric pipettes or 10- μL microsyringes.

⑧ 1 spatula or using automatic flotation machine.

⑨ 14 Pyrex glass pans.

⑩ 2 10 ml spray water bottles.

⑪Bucher funnels, Erlenmeyer flasks, filter paper (No. 42 Whitman), Tygon tubing, T-connectors, and vaccum line (building).

⑫ Crucibles, desiccators, and porcelain tray, tang, high temperature gloves.

⑬ Parr 5200 calorimeter.

⑭ Muffle Furnace.

(2) Personal Equipment

① Lab coat or old cloth.

② Personnel equipment (safety glasses, shoes, and respirator mask).

(3) Clean-up Equipment and Lab

① Liquid soap and paper towel.

② Floor brush or broom and vacuum cleaner.

4. Procedure

① Weight 200g coal sample with a top-loading balance.

② Put the coal sample into a 4L beaker. Add 300ml water, turn on the flotation machine at 1200 rpm and condition the sample for 10 minutes.

③ Pour the sample into the 2L flotation tank. Add water to half-cell. Add 16 drops collector and 10 drops frother. Turn on the flotation machine at 800 rpm and running for 3 minutes for pre-flotation.

④ Add water to the flotation tank until the water level reaches the guild line, turn on the switch of the timer and air valve, gather the froth on the glass collector and change the collector at the assigned time point, which are 15s, 30s, 45s, 60s, 90s, 120s, and 180s.

⑤ Label each glass collector with the time point, filtrate each collector and bake the froth coal and tailings for one hour, weight the coal sample for each time point and do the mass balance calculation.

5. Data Analysis and Report

① Follow the data analysis methods provided in the class to calculate yields, product ash and combustible material recovery at various flotation times by constructing three separate tables using spreadsheet software tools.

② Plot combustible material recovery % (cumulative) as a function of flotation time.

③ Plot product ash % (cumulative) as a function of flotation time.

④ Us the flotation experimental data and flotation rate model to construct the graphs using software tools, such as Excel, PSI-PLOT or custom flotation computer program. The graphs should include the experimental data points (using symbols) and calculated data (using solid line for predicted values) for the combustible material recovery and product ash as a function of flotation time.

⑤ For non-linear curve fitting, use the least-square method and Solver in Excel, or use PSI-PLOT non-linear curve fitting. The non-linear curve fitting will result in two parameter values for the flotation rate model. Use the known parameter values for the flotation model to generate the predicted data points for solid curve (connected by line).

Laboratory #5 - Thickener Design

1. Background

Thickening is a process used to remove excess liquid from the coal preparation plant black water and fine coal tailings slurry. The functions of the thickener include concentrating solids for dewatering, tailings thickening, clarified liquid or both, and countering current decantation in mineral industry. A thickener is a circular settling tank. The diluted tailings slurry or black water or mixture of both, as well as selected flocculant are fed into the center of the thickener. The overflow from the thickener is essentially clear liquid (thickener is also known as clarifier), and the underflow contains the solid materials plus some remaining liquid.

Flocculant is added to the feed slurry to hasten the sedimentation process. The flocculants increase the settling velocity of solids in a slurry by flocculating the fine particles into larger flocs. The typical flocculants include poly-acrylamine, and polyvalent cation.

2. Objective

The purpose of the experiment is for students to learn how to determine the thickener capacity such as area required from a solid settling curve obtained from a batch test. To meet this goal, students will:

① Perform the solid settling rate experiment in a graduated cylinder with flocculant solution.

② Plot interface versus time curve from the results of the experiment.

③ Determine the settling rate from the settling curve.

④ Design a thickener area for a given feed tonnage and underflow solid concentration.

3. Equipment Requirement

(1) Laboratory Equipment

① 2000 ml graduated cylinder.

② Stop watch.

③ 1-ml pipet or microsyringe.

④ Analytical balance.

(2) Personal Equipment

① Lab coat or old clothes.

② Personnel equipment (safety glasses, shoes, and respirator mask).

(3) Clean-up Equipment and Lab

① Liquid soap and paper towel.

② Floor brush or broom and vacuum cleaner.

4. Experimental Procedure

① The class will form several groups to work on the settling rates of solids with the addition of flocculent. Each group will work independently using fine coal tailings samples obtained from previous fine coal flotation test results.

② Prepare the stock solution of poly-arcrylaminde, a flocculant (original density of flocculant is 0.838 g/ml, which is diluted to 15,000 times).

③ In the 2000 ml graduated cylinder, prepare coal tailings slurry with 10% wt solid concentration, by using 200 gram tailings from the flotation of fine coal test.

④ The slurry is agitated gently and the stock solution of poly-arcylamide is added until the flocculation is started. The settling velocities of the solids are measured at various times, that are recorded at interphases of the graduated cylinder scales, including: 1800 ml, 1700 ml, 1600 ml, 1500 ml, 1400 ml, 1300 ml, 1200 ml, 1100 ml, 1000 ml, 900 ml, 800 ml, 700 ml, 600 ml 500 ml, 400 ml, and 300 ml.

5. Data Analysis

① Compile the settling rate experimental data in a table and plot the interphase height of solids in the column versus time in a chart using an Excel worksheet.

② Using the Kynch theory, and the example of data analysis for thickener design given in the computer lab, complete the thickener design procedures and submit your own group's settling velocity measured data for fine coal tailings slurry.

③ Design the thickener area requirement.

14.3 Exams

Exam 1 (Close books and notes)

(Note: Give the complete answers for the questions to receive full credits)

1. Briefly described the preparation of a "representative" coal sample, according to ASTM (American Standard Testing Methods) Standard for chemical analysis. (5%)

2. What components of the coal sample are involved in proximate analysis? (5%)

3. How do you determine the fixed carbon of coal? (5%)

4. The solid slurry was prepared and classified for froth flotation processes using classifying cyclone.

① Determine the mass balance, and the specific gravities of the slurries and percent solid concentrations. The specific gravity of solid and liquid are 4, and 1, respectively. The underflow solid mass (TPH) is 200% of overflow solid (TPH). Show the detailed calculations by giving the mass balance table of overflow, underflow, and feed. (8%)

② Fill-in the input and output legend boxes shown below. (4%)

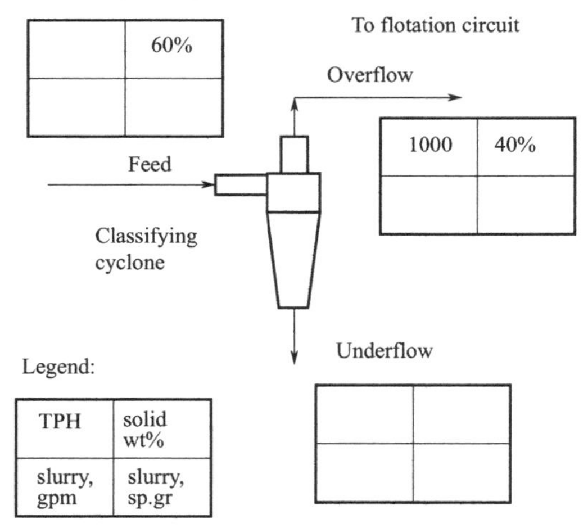

5. ① What is the principle of the communition process? (3%)

② State the actions or mechanisms of the communition process. (3%)

6. List the most commonly used three types of crushers in coal preparation plants ? (5%)

7. For coal size distribution, what mathematical equation can be used to describe the characteristics of the coal? Give the name, write the math equation, and define the parameters. (5%)

8. Name three types of most commonly used sizing equipment in coal preparation plants. (5%)

9. Describe the forms of sulfur in coal, and the forms of organic sulfur in coal. (5%)

10. ① What are the bases for classification of coal by rank in U.S. ? (3%)

② Name the coal ranks of coal in U.S. coal fields (give the major coal regions). (3%)

11. Describe how to determine (measure) the Hardgrove Grindability Index (HGI). (5%)

12. Why does it need coal cleaning? State the objectives. (6%)

13. Direct washability data for Alma Seam Coal is given in Table 14.1.

① Complete Table 14.1: Calculate the cumulative float wt% and ash%. (10%)

② Plot the cumulative float vs. ash curve. (5%)

③ Plot the cumulative float vs. specific gravity curve. (5%)

④ Calculate ±0.1 at specific gravity at 1.40,1.50 and 1.60, and then plot ± 0.1 near gravity curve. Is Alma seam coal easy or difficulty to clean. Explain the reason. (10%)

[Notes: draw your own graphs for plotting ② and ③]

Table 14.1 Washability Analysis for Alma Seam Coal for −6.12mm+1mm size fraction

Specific gravity interval	Direct		Cumulative Float	
	wt, %	Ash, %	wt, %	Ash,%
<1.30	28	6		
1.30~1.60	17	13		
1.60~1.80	23	32		
>1.80	32	55		

Exam 2 (Close books and notes)

Note: Give the complete answers for the questions to receive the full credits

1. Separation performance of coal cleaning circuits is particle size dependent. Give the particle size ranges, and the corresponding two most commonly used equipment names for each coal cleaning circuit. (12%)

① Coarse size range: 1)______, 2)______.

② Intermediate size range: 1)______, 2)______.

③ Fine size range: 1)______, 2)______.

④ Ultrafine size range: 1)______, 2)______.

2. ① Specific gravity-based separation processes are based on the settling velocity differences

between coal particles and mineral matter particles. Write the expression of "force balance" on a spherical shaped solid particle falling through the liquid. Define the notations used in the equation. (4%)

② Force balance on a solid sphere is related to the acceleration of the moving particle by F= ma where m is mass of a spherical particle; a is the acceleration of the particle in the fluid, which is expressed as a = dv/dt; where v is settling velocity and t is the time. Derive dv/dt of the solid particle settling in the fluid. Define the notations used for density and specific gravity. (4%)

③ Integrate the expression for dv/dt to obtain the settling velocity of particle, v, moving in the fluid. (4%)

3. Explain the benefit or advantage of using hinder-settling in dense medium, instead of using water only for free settling separation of coal. You must give an example to receive the full credit. (6%)

4. For ultrafine coal, froth flotation technology is applied in separation of ultrafine coal from mineral matters due to high separation efficiency and economic reasons. Give the principle of separating coal particles from mineral matter particles in fine coal flotation. (14%)

5. ① Name two commonly used collectors for flotation of fine coal. What is the purpose of using a collector? (8%)

② Name two commonly used frothers for flotation of fine coal. What is the purpose of using a frother? (8%)

6. In the United State, there are six ranked coal. Rank the flotability of those six coals (Use numerical sequential numbers, 1 being the most floatable and 6 being the least floatable). (12%)

7. Raw coal feed is characterized by washability data obtained from float-and-sink tests. To determine how difficulty is it to clean a raw coal, the amount of near gravity material in percent needs to be calculated at the specific gravity of separation, or cut point. (16%)

① Calculate the ±0.1 near gravity materials in percent for the washability data given in Table 14.2.

Table 14.2 Washability analysis for B seam coal for -2-in.+1/4-in. size fraction

Specific gravity interval	Direct		Cumulative Float	
	wt, %	Ash, %	wt, %	Ash, %
<1.30	25	6		
1.30~1.40	20	12		
1.40~1.50	17	20		
1.50~1.60	12	27		
1.60~1.70	11	30		
>1.70	15	42		

Calculate ±0.1 near gravity material at specific gravity of 1.50, and 1.60. Record the results in Table 14.3.

Table 14.3 ±0.1 near gravity material

Specific gravity	wt%, ±near gravity material	Ash, % (cumulative ash)
1.40		
1.50		

② How difficult is this coal for cleaning, if the coal is to be separated at 1.50 or 1.60 specific gravity? What type of coal cleaning unit operations can recommend to achieve the separation.

At 1.50,

At 1.60,

Exam 3 (Close books and notes)

1. Name the most commonly used unit operations for dewatering clean coal and refuse/tailings in coal preparation plants. (16%)

Clean Coal Refuse/Tailings

① Coarse dense medium cleaning circuit.

② Intermediate dense medium cleaning circuit.

③ Fine Spiral concentrator cleaning circuit.

④ Ultrafine coal flotation circuit.

2. Coal preparation plant produces black water with 1%~5% solid concentration. Name the unit operation which is used to treat the black water and state the purposes of treating black water: (6%)

3. Give one name of flocculant commonly used to flocculate fine particles for dewatering or thickening fine solid: Briefly explain this flocculant.(8%)

4. Name three major commercially available on-line coal quality analyzers:

What are the major applications in coal and/or power plant industries:(10%)

5. Fine coal flotation rate test results are given in Table 14.4. Calculate the feed ash%, and combustible material recovery% in clean coal (froth). (10%)

Table 14.4 Fine coal flotation test results and combustible material recovery

Time, sec	Weight, %	Ash, %	Ash Product, %	Combustible Material Recovery, %
15	33.37	7.11		
30	19.09	9.89		
60	20.10	13.62		

Time, sec	Weight, %	Ash, %	Ash Product, %	Combustible Material Recovery, %
120	12.80	18.25		
Tailing	14.65	77.34		
Feed	100.00			

6. The performance curve for a given gravity-based separator is given in Figure 14.1 (20%)

(a) Read SG50 (Cut point), and calculate Ep (Probable Error) and I (Im perfection). The float-and-sink test results with ash content are given in Table 14.5.

(b) Predict the coal cleaning results, including clean coal yield%, and clean coal ash%.

Table 14.5 Predicting coal cleaning results

Specific Gravity Interval	Direct (new raw coal)		Distribution factor reporting to product, %	Clean coal products		Clean coal product ash, %	Mean specific gravity
	wt, %	Ash, %		wt, %	product ash, %		
<1.30	20.00	7.00	100.00				
1.30~1.40	52.30	12.30	98.00				
1.40~1.50	11.40	23.80	90.00				
1.50~1.60	3.80	35.60	15.00				
1.60~1.70	1.90	41.80	7.00				
1.70~1.80	1.00	50.00	2.00				
>1.80	9.60	77.10	0.00				

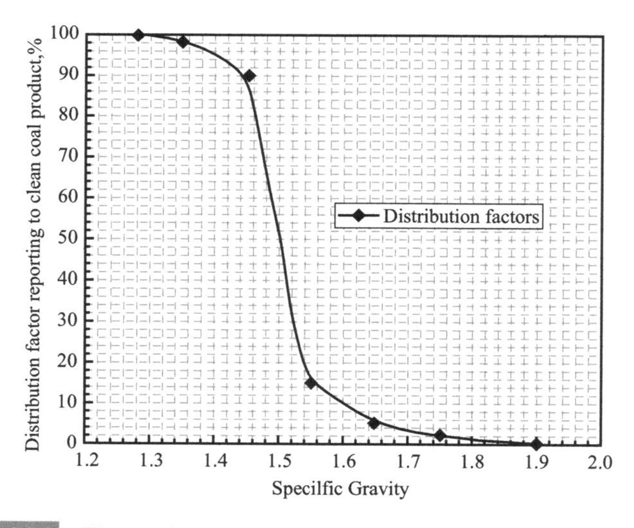

Figure 14.1 The performence curve for a given gravity-based separator

7. To calculate the filtration area for a plate & frame or filter press type of unit operation at constant pressure: (a) What mathematical formula can be used to design the unit operation? (b) Filtration test results are given in Table 14.6 with a total filtering area of l0 ft^2 and operating at given pressure drop. Assume the filter medium resistant is negligible. The slurry concentration is 4 lb/ft^3 and μ represents the viscosity of water. (15%)

i) To determine the compressibility of filter cake, evaluate the correlation of θ $(-\Delta p)/(V/A)$ versus V/A for the filtration test results by showing the correlation in a plot or chart.

ii) If the correlation (use plot from part i)) of the two test results is different, evaluate the correlation of the slopes and pressure drops by showing in a plot or chart.

iii) Calculate the compressibility coefficient, s, and empirical constant, α'.

Table 14.6 Filtration rate test results

Total volume of filtrate, V, ft^3	Time from state of filtration to θ (hr), at constant pressure drop of $-\Delta p$ (psi)	
	20	40
10	1.15	0.97
15	2.20	1.84

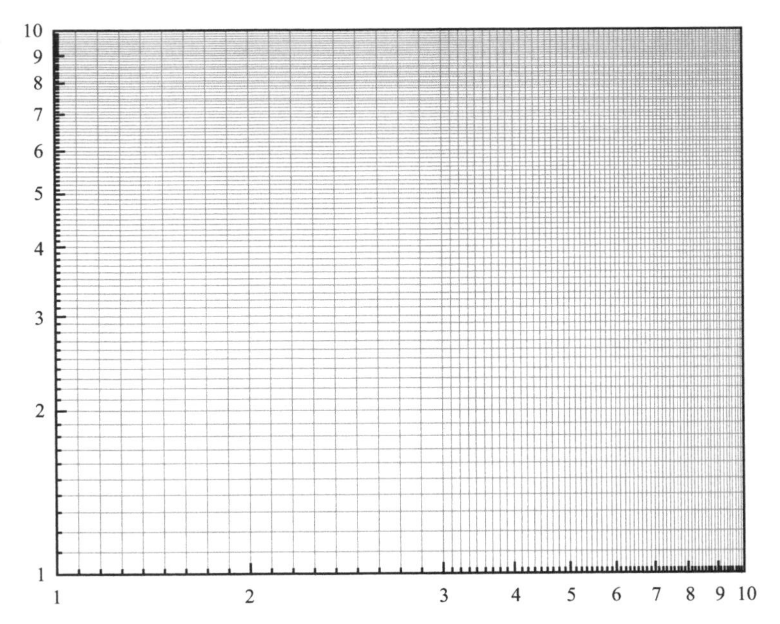

8. A batch test was conducted to determine the settling (sedimentation) velocity of slurry containing 100 kg solid/m^3. The test results are shown in Figure 14.2. Using the Kynch equations (shown in page 2), tabulate the settling (sedimentation) velocity, solid flux due to sedimentation as a function of concentration (use Table 14.7). Also estimate the area of tank required to give an underflow concentration of 400 kg/m^3 for a feed rate of 3m^3/min of slurry. (15%)

Table 14.7 Batch cylinder fine refuse particle settling test results

Height, cm	C, kg/m³	U_c, cm/min				

The area of thickener can be determined by:

$$A=\frac{Q_oC_o}{\Psi_T}=Q_oC_o\left[\frac{\frac{1}{C}-\frac{1}{C_u}}{u_c}\right]$$

Determine the maximum value of

$$\left[\frac{\frac{1}{C}-\frac{1}{C_u}}{u_c}\right]$$

Therefore, the minimum area of the thickener is

$$A=\frac{Q_oC_o}{\Psi_T}=Q_oC_o\left[\frac{\frac{1}{C}-\frac{1}{C_u}}{u_c}\right]\max$$

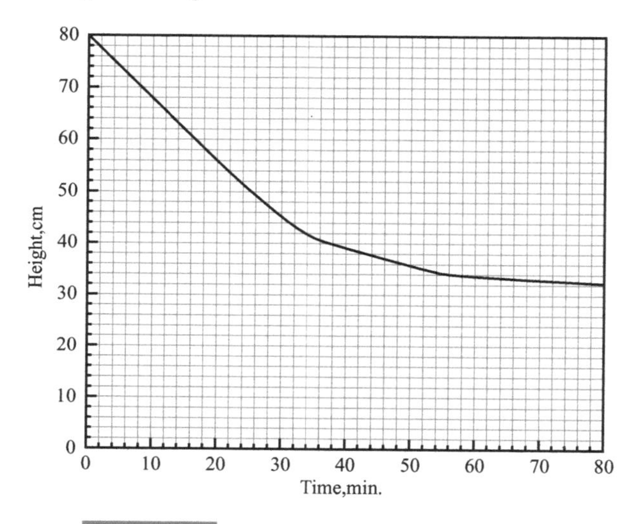

Figure 14.2 The test result height vs. time

where

A = area of tank

C_o = volumetric concentration of solid in feed, kg solid/m^3

C_u = concentration of slurry underflow, kg solid/m^3

C = concentration of slurry given sedimentation time, kg solid/m^3

Q_o = volumetric feed rate, m^3/min.

Ψ_T = the limiting total flux for a specified concentration C_u

Exam 4 (Open books)

1. Name the most commonly used unit operations for dewatering clean coal and refuse/tailing in coal preparation plants. (20%)

Cleaning Circuit	Clean Coal	Refuse or tailings
a. Coarse coal from dense medium vessel		
b. Intermediate coal from dense medium cyclone		
c. Fine coal from spiral		
d. Ultrafine coal from flotation		

2. To prepare metallurgical coal, if the market specification specifies to having 5% or less moisture content in the clean coal product, what would you do to meet this specification? (5%)

3. ① Coal preparation plant produces black water with 1%~5% solid concentration. Name the unit operation which can be used to treat the black water.

② State the purpose of treating the black water.

③ briefly state the operating principle of separation of solid-liquid in this unit operation. (12%)

4. ① Give one name of the most commonly used chemical to flocculate the fine particles. (b) Briefly explain the function of this flocculant.

② Name two major functional groups in this flocculant.

③ There are two primary functional groups in this flocculant. Name the functional groups and their function. (9%)

5. The vacuum disc filter is commonly used as continuous dewatering equipment in coal preparation plants.

① Briefly describe the operating principle of this unit's operation.

② Write the mathematical expression for the time per revolution of the disc, and the optimal operation time per revolution ranges, and the pressure (drop) applied:

③ The cake surface moisture depends upon the minus 200 mesh particle fraction in the dry cake. If the dry cake contains 35% of the size fraction, what is the % surface moisture by wt? (9%)

6. For centrifugal filtration, the pressure (driving force) is due to centrifugal action. Write the mathematical formula for this action. (5%)

7. ① To calculate the filtering area for a plate & frame or filter press type of unit operation at constant pressure, What mathematical equations can be used to design the unit operation?

② Filtering experiment results are given in Table 14.8 with a total filtering area of 10 ft^2 and operating at a given pressure drop. Assume the filter medium resistant is negligible. The slurry concentration is 4 lb/ft^3, where μ is viscosity of liquid.

i) To determine the compressibility of filter cake, evaluate the correlation of $\theta(-\Delta p)/(V/A)$ versus V/A for the filtration test results by showing the plot or chart.

ii) If the correlation [see part i)] of the two test results is different, evaluate the correlation of the slopes and pressure drops by showing them in a plot or chart. Then, calculate the compressibility coefficient, s, and empirical constant of α'. (20%)

Table 14.8 Filtering experiment results

Total volume of filtrate, V, ft^3	Time from start of filtration to θ (hr), at constant pressure drop of $(-\Delta p)$ (psi)	
	20	40
10	1.15	0.97
15	2.20	1.84

8. A batch test was conducted to determine the settling (sedimentation) velocity of slurry containing 100 kg solid/m^3. The test results are shown in Figure 14.3. Using the Kynch equations, tabulate the settling (sedimentation) velocity, solid flux due to sedimentation as a function of concentration (use Table 14.9). Also estimate the area of tank required to give an underflow concentration of 400 kg/m^3 for a feed rate of 3m^3/min of slurry. (20%)

Table 14.9 Batch test results

Height, cm	C	u_c			

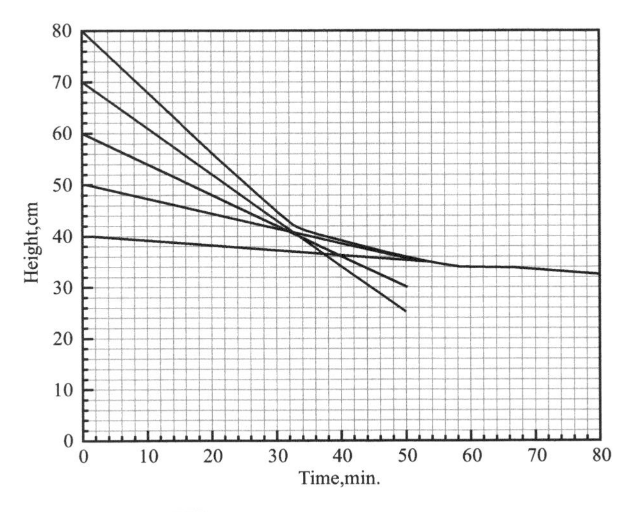

Figure 14.3 The test results height vs. time

Area of thickener:

$$A=\frac{Q_oC_o}{\Psi_{TL}}=Q_oC_o\left[\frac{\frac{1}{C_L}-\frac{1}{C_u}}{u_{cL}}\right]$$

Determine the maximum value of

$$\left[\frac{\frac{1}{C_L}-\frac{1}{C_u}}{u_{cL}}\right]$$

Therefore the minimum area of the thicker is

$$A=\frac{Q_oC_o}{\Psi_{TL}}=Q_oC_o\left[\frac{\frac{1}{C}-\frac{1}{C_u}}{u_c}\right]_{max}$$

where

A=area of tank;

C_o=volumetric concentration of solid in feed, kg solid/m^3;

C_u=concentration of slurry underflow, kg solid/m^3;

C_L=concentration of slurry at given sedimentation time, kg solid/m^3;

Q_o=volumetric feed rate, m^3/min;

Ψ_{TL} =the limiting total flux for a specified concentration C_u.

Final Exam (Open book)

1. State three reasons for coal cleaning. (5%)

2. What is to be determined for the proximate analysis and ultimate analysis of coal? (5%)

3. What forms of sulfur exist in coal? Which form of sulfur can be removed and minimized by using the physical coal separation method? If a coal sample has 2.4% of total sulfur, and a heat content of 13,400 Btu/lb, calculate the pound of sulfur dioxide per million Btu emitted from a boiler. (5%)

4. Size analysis of a coal sample is shown as follows (5%):

US sieve	Weight	wt, %	Cumulative wt, %
+1 in.	55.2		
−1+1/4 in.	61.3		
−1/4 in.+No.16	75.1		
−No.16 + No.50	44.8		
−No.50 + No.100	38.4		
−No.100	21.3		
Feed			

Calculate the size distribution in wt%, and cumulative wt%. Rosin-Rammler Equation can be used to describe the coal particle size distribution of the coal sample. Derive the Rosin-Rammler equation for this coal sample, using the Rosin-Rammler graph (read the graph, and express the final equation in "mm" unit).

5. Define the Hardgrove Grindability Index (HGI). What is the application of HGI? (5%)

6. What's the major cause of energy consumption in size reduction? What are the three laws for expressing the relationship between energy requirement and particle size for size reduction. (5%)

7. Name two most commonly used crushers in coal preparation plant. (note: no credit for the uncommon unit operations named). (3%)

8. A set of washability data for a coal sample is given as follows: (10%).

Specific Gravity Interval	Direct		Cumulative float	
	wt, %	Ash, %	wt, %	Ash, %
<1.3	40	7.5		
1.3~1.6	30	16.2		
1.6~1.8	20	33.7		
>1.8	10	55		

① Complete the above table.

② Plot cumulative float vs. ash%.

③ Plot cumulative float vs. specific gravity curve.

④ Read the curve (in ②) to find the theoretical yield of the clean coal containing 12% ash.

⑤ Read the curve (in ③) to find the theoretical specific gravity of separation for producing a clean coal product of 12% ash content.

⑥ By reading the curves, calculate ±0.1 near gravity material at the theoretical specific gravity of separation from ⑤ .

9. A flotation of fine coal test was conducted. The sample weight% and ash% are shown in the following table. Calculate the (Cumulative) Combustible Material Recovery% (CMR%) by completing the blank columns in the table. (10%)

Time, s	Direct, %		Cumulative, %		Direct CMR, %	Cumulative CMR, %
	wt	Ash	wt	Ash		
15	51	7				
30	23	9				
60	7	12				
120	3	14				
180	2	16				
300	1	18				
Refuse	13	40				

10. Answer the following questions: (6%)

① Define the major ash components in coal ash.

② What is the major impact on the boiler equipment from those ash compositions during combustion?

③ What are the criteria used to express the degree of severity affected by those compositions?

11. Calculate the mass balances for feed, overflow and underflow streams of a sizing cyclone. Fill in the blanks for each stream in the table. (10%)

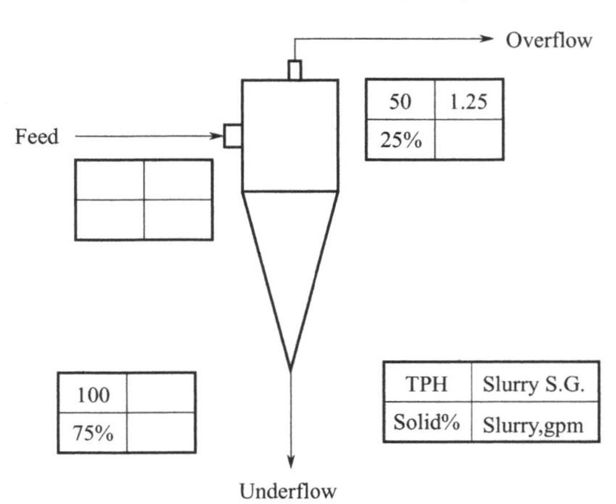

A. Determine the mass balance, S.G. of solid and slurry, and flow rate of overflow				
	TPH	S.G.	gpm	
Solid	50			
Water				
Slurry		1.25		
Solid% by wt	25	—		Solid% by vol.
B. Determine the mass balance, S.G. of solid and slurry, and flow rate of underflow				
Solid	100			
Water				
Slurry				
Solid% by wt	75	—		Solid% by vol.
C. Determine the mass balance, S.G. of solid and slurry, and flow rate of feed				
Solid				
Water				
Slurry				
Solid% by wt		—		Solid% by vol.

12. A coal preparation plant has a water circuit which flows at 5000 gpm of coal fine slurry with a solid to liquid ratio of 1∶50. In considering the use of a thickener to prevent slime build-up, settling tests were conducted to determine the settling rates. The density of the solid is 1.6. The solid to liquid ratio of underflow of the thickener is 1∶2. The test results are given in the following table. Determine the proper diameter of the thickener. (10%)

Solid to liquid ratio	Settling velocity, ft/hr
1 : 30	1
1 : 20	0.5
1 : 10	0.2
1 : 5	0.12
1 : 3	0.08

13. For a specific gravity based coal cleaning separator, the separation performance of the clean coal separator can be described by the distribution factors reporting to clean coal stream as shown in the following figure. The washability data of a new raw coal is given in the table below. (10%)

① Predict the clean coal yield%, and ash% of clean coal to be produced.

② What is the specific gravity of separation?

③ What is the probable error?

④ What is imperfection?

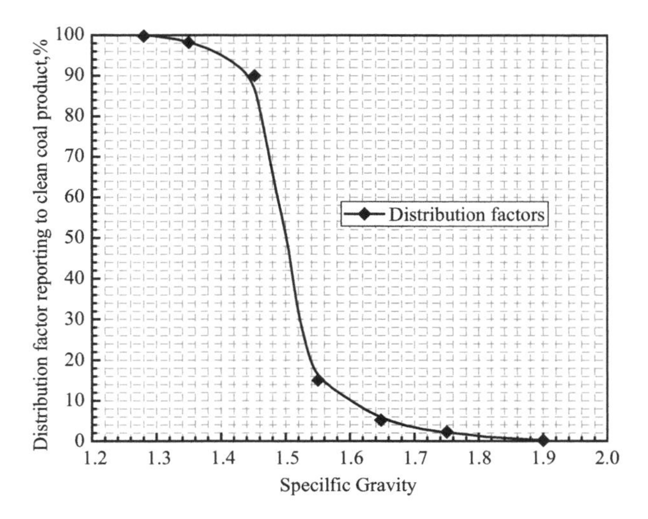

Specific Gravity Interval	Direct		Distribution Factors, %	wt, %	Ash, %	
	wt, %	Ash, %				
<1.30	23	7				1.28
1.30~1.40	10	9				
1.40~1.50	18	11.2				
1.50~1.60	24	18.6				
1.60~1.70	7	40.1				
1.70~1.80	3	55.0				
>1.80	15	68.5				2.0

14. Name the "most commonly used" dewatering unit operations in coal preparation plants. (note: no credit for naming an uncommonly used unit operation in coal preparation plant). (4%)

For dense medium vessel products:

For dense medium cyclone products:

For spiral concentrator products:

For froth flotation products:

15. Describe the procedures for determining the filter area (A) requirement for a given feed slurry at constant filtration pressure and low Reynolds number. Use the filtration equations and experimental tests to explain the procedures. (4%)

16.(Extra credit problem, 25%)

This is a good example of the mass balance problem in the processing plants. The Cleaner flotation can be used to improve the quality of the final product, and the scavenger flotation used to improve the final CMR. The underflow of the cleaner flotation and overflow of the Scavenger flotation are combined as middlings, and returned to Rougher flotation. The flowsheet of the example is shown in the figure below. The ash analysis of each input and output is given below. Calculate the yield% and CMR% of locations 2, 4, 5 and 6.

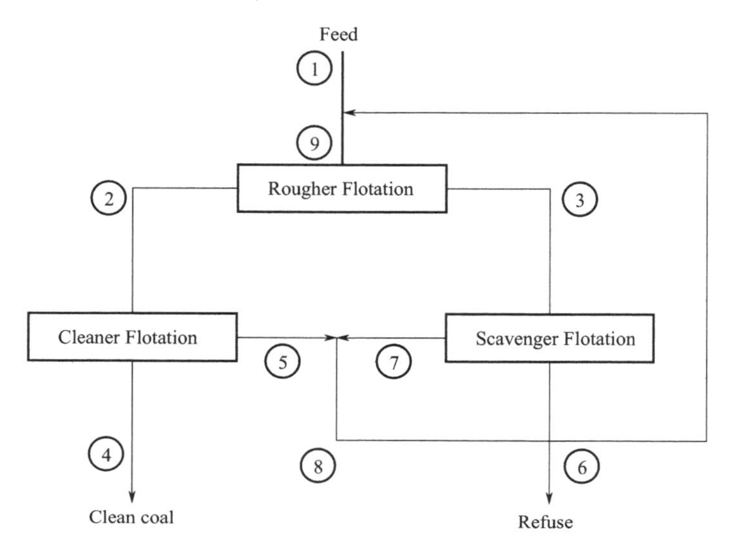

No.	Ash, %	Yield, %	CMR, %	
1	32.0			
2	13.0			
3	40.0			
4	10.0			
5	35.0			
6	68.0			
7	25.0			
8				
9				

References

[1] 911 Metallurgy Corp. Filtration -Leaf Filter Testing. Retrieved from https://www.911metallurgist.com/blog/filtration-leaf-filter-testing.

[2] 911 Metallurgy Corp. Vacuum Disk Filter. Retrieved from https://www.911metallurgist.com/equipment/rotary-drum-filters/.

[3] ABC Machinery. Coal Slime Dryer. Retrieved from http://www.rotary-air-dryer.com/Industrial-Dryer-Machine/Coal-Slime-Dryer.html.

[4] Ambrós, W.M. (2020). Jigging: A Review of Fundamentals and Future Directions, (https://www.researchgate.net/publication/345686319).

[5] ASTM D1857-87 (1994) Standard Test Method for Fusibility of Coal and Coke Ash. ASTM International, DOI: 10.1520/D1857-87R94 www.astm.org.

[6] ASTM D2013-68 (2007) Standard Practice for Preparing Coal Samples for Analysis. ASTM International, www.astm.org.

[7] ASTM D2014-96a (1997) Standard Test Method for Expansion or Contraction of Coal by the Sole-Heated Oven. ASTM International, www.astm.org.

[8] ASTM D2492-90 (1998) Standard Test Method for Forms of Sulfur in Coal. ASTM International, DOI: 10.1520/D2492-90R98 www.astm.org.

[9] ASTM D2639 (2010) Standard Test Method for Plastic Properties of Coal by the Constant-Torque Gieseler Plastometer. ASTM International, DOI: 10.1520/D2639-08 www.astm.org.

[10] ASTM D2961-87 (1987) Standard Test Method for Total Moisture in Coal Reduced to NO. 8 (2.36-mm) Top Sieve Size (Limited-Purpose Method) (E1-1990). ASTM International, www.astm.org.

[11] ASTM D3173-87 (1996) Standard Test Method for Moisture in the Analysis Sample of Coal and Coke. ASTM International, ICS Code: 75.160.10 www.astm.org.

[12] ASTM D3174-00 (2017) Standard Test Method for Ash in the Analysis Sample of Coal and Coke from Coal. ASTM International, DOI: 10.1520/D3174-00 www.astm.org.

[13] ASTM D3175-89A (1989) Standard Test Method for Volatile Matter in the Analysis Sample of Coal and Coke. ASTM International, www.astm.org.

[14] ASTM D3176-89 (2002) Standard Practice for Ultimate Analysis of Coal and Coke. ASTM International, DOI: 10.1520/D3176-89R02 www.astm.org.

[15] ASTM D3177-89 (2002) Standard Test Methods for Total Sulfur in the Analysis Sample of Coal and Coke. ASTM International, DOI: 10.1520/D3177-89R02 www.astm.org.

[16] ASTM D3302-07a (2010) Standard Test Method for Total Moisture in Coal. ASTM International, DOI:

10.1520/D3302-07A www.astm.org.

[17] ASTM D346 (2019) Standard Practice for Collection and Preparation of Coke Samples for Laboratory Analysis. ASTM International, DOI: 10.1520/D0346_D0346M-11R19E01 www.astm.org.

[18] ASTM D388 (1998) Standard Classification of Coals by Rank. ASTM International, www.astm.org.

[19] ASTM D409 (2016) Standard Test Method for Grindability of Coal by the Hardgrove-Machine Method. ASTM International, www.astm.org.

[20] ASTM D4371-06 (2012) Standard Test Method for Determining the Washability Characteristics of Coal. ASTM International, DOI: 10.1520/D4371-06.

[21] ASTM D440-86 (2002) Standard Test Method of Drop Shatter Test for Coal. ASTM International.

[22] ASTM D5142-90 (1990) Standard Test Methods for Proximate Analysis of the Analysis Sample of Coal and Coke by Instrumental Procedures. ASTM International, global.ihs.com.

[23] ASTM D5865-99a (2021) Standard Test Method for Gross Calorific Value of Coal and Coke. www.astm.org.

[24] ASTM E11 (2022) Standard Specification for Woven Wire Test Sieve Cloth and Test Sieves. ASTM International, www.astm.org.

[25] Averitt, P. (1975). Coal Resources of the United States, January 1, 1974. U.S. Geological Survey, Bulletin 1412; Washington D.C.: U.S. Department of Interior; 131 pp.

[26] Ayat, M. G. and Scott, B. C., (1986). Spiral separators versus concentrating tables-A comparison for the coal preparation industry, SME preprint 86-320.

[27] Coe, H. S. and G. H., Clevenger (1916). Methods for Determining the Capacities of Slime-Settling Tanks, Trans., Amer. Inst. Min. Engrs., 60, 356-384.

[28] EIA (2006). Annual Energy Review 2005. Report No. DOE/EIA-0384(2005). Washington, D.C.: U.S. Department of Energy. Available online at http://www.eia.doe.gov/emeu/aer/contents.html; accessed April 2007.Levine, D. G., Schlosberg, R. H., and Silbernagel, B. G. (1982). Understanding the Chemistry and Physics of Coal Structure: A review. Proc. Natl. Acad. Sci. USA. 79:3365-3370.

[29] Enviro-Clear Company, Inc. High Capacity Clarifier/Thickeners. Retrieved from http://enviro-clear.com/clarifier-thickener/high-cap.

[30] FLSmidth Krebs Manifolds. Retrieved from https://www.flsmidth.com/en-gb/products/centrifugation-and-classification/krebs-manifolds.

[31] Hyman, D. M. (2002). Dense media Cyclone Optimization, (https://www.osti.gov/servlets/purl/802860) Intime Associates. Retrieved from https://www.intimeassociates.com/services-hydrocyclones.html#.

[32] Kentucky Geological Survey, University of Kentucky. Float-Sink (Washability) Test. Retrieved from https://www.uky.edu/KGS/coal/coal-analyses-float-sink.php.

[33] King, R.P. (2003). Introduction to Practical Fluid Flow. 1st edition. Butterworth-Heinemann, 208 pages. Klimas, C., etc. (2009). An Ecosystem Restoration Model for the Mississippi Alluvial Valley Based on Geomorphology, Soils, and Hydrology.

[34] Kujawa, C. (2011). Cycloning of Tailing for the Production of Sand as TSF Construction Material, Proceedings Tailings and Mine Waste 2011, Vancouver, BC, November 6 to 9.

[35] Kynch, G. J. (1952). A Theory of Sedimentation. Trans., Faraday Soc., 48, 166-176.

Leonard, J.W. (1991), editor. Coal Preparation. 5th edition. Society for Mining, Metallurgy, and Exploration, Inc. Littleton, Colorado.

[36] Michaud, D. (2021). https://www.911metallurgist.com/hydrocyclone-coal-washing-equipment/.

[37] Micronics Engineered Filtration Group, Inc. Rotary Vacuum Disc Filter Basics. Retrieved from https://www.micronicsinc.com/filtration-news/rotary-vacuum-disc-filter-basics/.

[38] Miltech Energy Service, Inc. Coal Preparation Plant Assessment and Design. Retrieved from https://www.miltechenergy.com/coal-preparation-plant-assessment-design.php.

[39] Mohanty, M, B. Zhang, H. Akbari (2008). Evaluation of FGX dry separator for cleaning Illinois Basin coal, project report.

[40] Mohanty, M. (2007). Genetic algorithms — A novel technique to optimize coal preparation plants, International Journal of Mineral Processing 84(1-4) p. 133-143.

[41] Morrill, J., P. Sampat, P. Personius (2022). Safety First Guidelines for Responsible Mine Tailings Management V2.0. Retrieved from https://earthworks.org/resources/safety-first/.

[42] Nagaraj, D. R. (2005). Minerals Recovery and Processing, (https://www.researchgate.net/figure/Gravity-dense-medium-drum-separator-where-is-the-float-and-the-sink-4_fig6_228013236).

[43] National Academies of Sciences, Engineering, and Medicine (2007). Coal: Research and Development to Support National Energy Policy. Washington, DC: The National Academies Press. Retrieved from https://doi.org/10.17226/11977.

[44] National Academies Press (2007). Coal: Research and Development to Support National Energy Policy. Retrieved from https://nap.nationalacademies.org/catalog/11977/coal-research-and-development-to-support-national-energy-policy.

[45] National Coal Council (2006). Coal: America's Energy Future, Volumes I and II. Washington, D.C.

Port of Pittsburgh Commission (2021). The Port of Pittsburgh: Impact, Opportunities, and Challenges, Final Report, February 17, 2021.

[46] Schweinfurth, S. P. (2002). Coal-A complex natural resources: An overview of factors affecting coal quality and use in the United States, US Geological Survey Circular.

[47] Tangel, O. F., and Brison, R.J., 1956. Bibliography of the Liquid-solid Cyclone, Journal of the Southern African Institute of Mining and Metallurgy, Volume 56, Issue 8, pps 299-302.

[48] Tewalt, S.J., Bragg, L.J., and Finkelman, R.B. (2001). Mercury in U.S. Coal—Abundance, Distribution and Modes of Occurrence. U.S. Geological Survey Fact Sheet FS-095-01.

[49] Tully, J. (1996). Coal Fields of the Conterminous United States, USGS Open File Report of 96-92.

[50] U.S. Department of Energy (1999). Carbon Sequestration Research and Development, DOE/SC/FE-1, U.S. Government Printing Office, Washington, DC.

[51] U.S. Energy Information Administration (2023). Coal Explained-Use of Coal. Retrieved from https://www.eia.gov/energyexplained/coal/use-of-coal.php.

[52] Vick, S. G. (1990). Planning, Design, and Analysis of Tailings Dams, BiTech Publishers Ltd. Vancouver, B.C. Canada.

[53] vividmaps.com (2023). The Mississippi River Basin. Retrieved from https://vividmaps.com/Mississippi-river/.

[54] Weir Global Screen Equipment. Retrieved from https://www.global.weir/product-catalogue/screening-equipment/enduron-banana-screens.

[55] Wells, S. A. (1998). Filtration Modeling of a Plate-and-Frame Press. Fluid/Particle Separation journal, 11(2), 152-165.

[56] West Virginia Geological and Economic Survey, Morgantown, WV (1980). Retrieved from https://www.wvgs.wvnet.edu.

[57] West Virginia Office of Energy. retrieved from https://www.energywv.org/wv-energy-profile/coal.

[58] WesTech Engineering, LLC. Horizontal Belt Filter. Retrieved from https://www.westech-inc.com/products/horizontal-belt-filter.

[59] Wills, B. A. (2006). Mineral Processing Technology-An Introduction to Practical Aspects of Ore Treatment and Mineral Recovery, (7th edition).

[60] Wills, B. A. (2016). Mineral Processing Technology-An Introduction to Practical Aspects of Ore Treatment and Mineral Recovery, (8th edition).

[61] Wills, B.A. and J. A. Finch (2015). Wills' Mineral Processing Technology: An Introduction to the Practical Aspects of Ore Treatment and Mineral Recovery, 8th Edition, Elsevier Ltd.

Subject Index